21世纪高等院校移动开发人才培养规划教材
21Shiji Gaodeng Yuanxiao Yidong Kaifa Rencai Peiyang Guihua Jiaocai

Android项目开发入门教程

张伟华 主编　朱东 伊雯雯 副主编　李金友 高乐 参编

Android Inventor Tutorial

人民邮电出版社

北京

图书在版编目（CIP）数据

Android项目开发入门教程 / 张伟华主编. -- 北京：
人民邮电出版社，2015.12
　21世纪高等院校移动开发人才培养规划教材
　ISBN 978-7-115-37927-6

Ⅰ. ①A… Ⅱ. ①张… Ⅲ. ①移动终端－应用程序－
程序设计－高等学校－教材 Ⅳ. ①TN929.53

中国版本图书馆CIP数据核字(2015)第123814号

内　容　提　要

本书精选 4 个典型项目贯穿全书，通过任务分解的方式，逐步学习各个知识点。第一个项目为开发准备，主要介绍 Android 开发环境搭建和 Android 程序的创建步骤。第二个项目为"我爱记单词"，融合了移动商务所需的各项技术，主要介绍界面设计、网络互连、数据管理等内容。第三个项目为数独游戏，主要介绍游戏编程，包括游戏编程所需的技术如图形用户界面、资源、Intent、Activity、Servers、Broadcast、动态图形绘制、音频文件控制等内容。第四个项目为手机定位应用，主要介绍短信的互发与拦截、GPS 定位与地图编程、传感器的编程等手机终端应用编程技术。

本书既可作为院校计算机专业的教科书，又可作为从事 Android 应用开发工程技术人员的入门参考书。

◆ 主　　编　张伟华
　　副主编　朱　东　伊雯雯
　　参　　编　李金友　高　乐
　　责任编辑　范博涛
　　责任印制　杨林杰

◆ 人民邮电出版社出版发行　北京市丰台区成寿寺路 11 号
　邮编　100164　电子邮件　315@ptpress.com.cn
　网址　http://www.ptpress.com.cn
　北京鑫正大印刷有限公司印刷

◆ 开本：787×1092　1/16
　印张：14.75　　　　　　2015 年 12 月第 1 版
　字数：368 千字　　　　　2015 年 12 月北京第 1 次印刷

定价：35.00 元
读者服务热线：(010)81055256　印装质量热线：(010)81055316
反盗版热线：(010)81055315

前言 PREFACE

中国已经是全球手机用户最多的国家，随着 3G 和 4G 网络的广泛使用，移动互联网的迅猛发展，智能手机日益普及，手机移动开发已成为一个高速发展的领域。Android 作为一个开放的平台，适用各种移动终端，用户数量日益增长。基于 Android 技术的软件开发应用前景非常广阔，高校学生及广大软件开发者认识到这个趋势，踊跃加入到学习 Android 的行列中来。因此，编写一本易用于教学和学习的教材显得非常必要。

目前，市面上 Android 技术的书籍对于知识的介绍一般都是按章节来阐述，案例的处理一般都只针对一个知识点来展开介绍，缺乏对一个完整项目实施过程的介绍。当前教育普遍推行基于工作过程的任务驱动教学模式，市场上这类教材比较少，难以满足教学需求。本书基于 Android 学习和教学的需要进行编写，按照任务驱动的方式组织学习内容，按照应用需求分类将 Android 主要应用技术依托 4 个项目进行介绍，每个项目以任务的方式分解完成，在任务介绍过程中，依次按照任务目标、任务内容、任务分析、实现过程、技术要点、技能训练、任务拓展七部分展开。在任务实施过程中加强动手实践的能力，在技术要点中介绍任务中所需要的技术，并在技能训练和任务拓展环节辅以任务训练，强化对内容的理解。

本书由具有多年从事 Java、Android 技术开发和教学经验的教师编写而成。苏州工业职业技术学院的张伟华老师任主编，朱东、伊雯雯老师任副主编，企业工程师李金友、高乐对书的撰写提供了许多案例和宝贵意见，在此表示衷心感谢！

由于编者水平有限，书中难免存在疏漏和不足，恳请同行专家和读者能给予批评和指正。

编 者
2015 年 9 月

目 录 CONTENTS

项目一　开发准备　1

任务 1.1　开发环境搭建　2
　任务描述　2
　任务目标　2
　任务分析　2
　知识要点　3
　任务实现　6
任务 1.2　第一个 Android 应用程序开发 12
　任务描述　12
　任务目标　12
　任务分析　12
　知识要点　13
　任务实现　21
　技能训练　26

项目二　我爱记单词　27

任务 2.1　用户登录界面设计　29
　任务描述　29
　任务目标　29
　任务分析　29
　知识要点　30
　任务实现　37
　技能训练　43
任务 2.2　系统退出功能设计　44
　任务描述　44
　任务目标　45
　任务分析　45
　知识要点　45
　任务实现　47
　技能训练　49
任务 2.3　用户注册界面设计　49
　任务描述　49
　任务目标　50
　任务分析　50
　知识要点　51
　任务实现　56
任务 2.4　用户访问信息存取　64
　任务描述　64
　任务目标　65
　任务分析　65
　知识要点　65
　任务实现　67
　技能训练　71
任务 2.5　单词存取　71
　任务描述　71
　任务目标　71
　任务分析　72
　知识要点　72
　任务实现　75
　技能训练　83
任务 2.6　单词共享　85
　任务描述　85
　任务目标　85
　知识要点　86
　任务分析　90
　任务实现　90
　拓展学习　98
任务 2.7　用户信息网络传输　99
　任务描述　99
　任务目标　99
　任务分析　99
　知识要点　100
　任务实现　102
　技能训练　113

任务 2.8 单词网络下载	114	任务分析	115
任务描述	114	知识要点	115
任务目标	115	任务实现	116

项目三　数独游戏　131

任务 3.1 数独游戏界面设计	132	任务分析	149
任务描述	132	知识要点	149
任务目标	133	任务实现	149
任务分析	133	任务 3.4 游戏背景音乐设置	156
知识要点	133	任务描述	156
任务实现	133	任务目标	156
任务 3.2 九宫格界面绘制	140	任务分析	156
任务描述	140	知识要点	157
任务目标	140	任务实现	158
任务分析	141	任务拓展	162
知识要点	141	任务 3.5 继续游戏功能实现	166
任务实现	142	任务描述	166
技能训练	147	任务目标	167
任务 3.3 数字键盘设计与实现	148	任务分析	167
任务描述	148	知识要点	167
任务目标	148	任务实现	168

项目四　手机定位应用　174

任务 4.1 界面与数据层设计	175	知识要点	185
任务描述	175	任务实现	189
任务目标	176	任务拓展	195
任务分析	176	任务 4.3 地图显示联系人位置	203
任务实现	176	任务描述	203
技能训练	184	任务目标	203
任务 4.2 短信发送与接收处理	184	任务分析	204
任务描述	184	知识要点	204
任务目标	184	任务实现	208
任务分析	184	任务拓展	219

PART 1 项目一 开发准备

项目情境

"工欲善其事必先利其器",在开始 Android 开发之旅之前,要先搭建开发环境。作为一个 Android 应用程序开发人员,掌握 Android 开发环境的配置是必须的,只有掌握了最基本的环境配置,才能进行后续的项目开发。在本项目中我们要完成 Android 应用程序开发环境搭建,然后完成第一个 Android 应用程序。

学习目标

- ☑ 能够下载 Android 的开发工具包。
- ☑ 能够搭建 Android 开发环境。
- ☑ 能够创建 Android 应用程序。
- ☑ 掌握 Android 应用程序框架。

工作任务

任务名称
任务 1.1 开发环境搭建
任务 1.2 第一个 Android 应用程序开发

任务 1.1 开发环境搭建

任务描述

在进行 Android 开发之前,需要搭建相应的开发环境。本任务主要实现 Android 开发环境的搭建,包括 JDK 的安装与配置、ADT 和 AVD 的配置、Android SDK 的配置。

名词解析

Java 开发工具包(Java Development Kit,JDK):是 Java 开发所必需的开发包。

Android 开发工具(Android Developer Tools,ADT):是集成在 Eclipse 上开发调试 Android 应用程序的工具。

Eclipse:是基于 Java 的可扩展开发平台。Android 推荐使用 Eclipse 来开发 Android 应用,并为它提供了专门的插件 ADT。

Android 软件开发工具包(Android Software Development Kit,Android SDK):是 Android 专属的软件开发工具包。

Android 系统模拟运行设备(Android Virtual Device,AVD):用于在电脑上模拟手机进行 Android 应用程序运行的设备。

Java 应用程序接口(Application Programming Interface,API):是 Java 开发工具中预先定义的函数,为软件开发人员提供访问某软件或硬件例程代码。

任务目标

① 了解 Android 的历史和版本。
② 了解 Android 的系统架构。
③ 知道如何下载 Android 开发所需的工具包。
④ 能安装并配置 JDK。
⑤ 能配置 ADT,并能使用 ADT 进行 Android 应用程序开发。
⑥ 能配置 Android SDK,并能进行 Android SDK 的升级管理。
⑦ 会配置 AVD,并能使用 AVD 进行 Android 应用程序运行调试。

任务分析

本书所用的 Android 开发环境的工具包主要包括 JDK、Android SDK、Eclipse 和 ADT 等软件包。这些开发工具包都可以通过网络下载,具体实现过程:

① 下载工具包;
② 安装配置 JDK;
③ 安装配置 ADT;
④ 配置 Android SDK;

⑤ 配置并启动 AVD。

知识要点

1. Android 的由来

Android 是一种基于 Linux 内核的开放源代码的操作系统，主要应用于移动设备，如智能手机和平板电脑，由 Google 公司和开放手机联盟领导及开发。在中国较多人将其翻译为"安卓"。

2003 年，Andy Rubin 等人创建 Android 公司；2005 年 Google 公司收购 Android 公司后，继续开发运营 Android 系统；2008 年 Google 公司推出了 Android 的最早版本 Android 1.0；2009 年 Google 公司推出了 Android 1.5。从这个版本开始，Android 的后续版本均用一个甜品的名称来命名。随着后续发展，越来越多的"甜品"（Android 版本）被 Google 公司陆续推出，让我们来认识一下这些"甜品"吧。

下面依次介绍 Android 版本。

2009 年 4 月发布的 Android 1.5：Cupcake（纸杯蛋糕）。

2009 年 9 月发布的 Android 1.6：Donut（甜甜圈）。

2010 年 1 月发布的 Android 2.1：Éclair（巧克力泡芙）。

2010 年 5 月发布的 Android 2.2：Froyo（冷冻酸奶）。

2010 年 12 月发布的 Android 2.3：Gingerbread（姜饼）。

2011 年 2 月发布的 Android 3.0/3.1：Honeycomb（峰巢）。

2011 年 10 月发布的 Android 4.0：Ice Cream Sandwich（冰淇淋三明治）。

2012 年 6 月发布的 Android 4.1：Jelly Bean（果冻豆）。

2013 年 9 月发布的 Android 4.4：KitKat（奇巧巧克力）。

2014 年 6 月 Google 公司发布了最新的 Android L，即 Android5.0 系统。

各版本的 Logo 如图 1-1 所示。

图 1-1 Android 版本

2．Android 系统架构

Android 系统架构从软件分层角度来看，可分为应用程序层（Application）、应用框架层（Application Framework）、系统库（Libraies）、Android 运行时（Android Runtime）、Linux 内核层（Linux Kernel）5 个部分。Android 系统架构如图 1-2 所示。

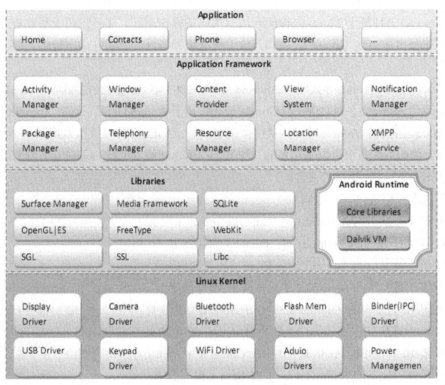

图 1-2　Android 系统架构

（1）应用程序层

应用程序层（Application）包含在 Android 设备上运行的所有应用，它们是 Android 系统中直接面向用户的部分。

Android 平台不仅仅是操作系统，也包含了许多应用程序，如 SMS 短信客户端程序、电话拨号程序、图片浏览器、Web 浏览器等应用程序。这些应用程序是用 Java 语言编写的，且这些应用程序都是可以被其他应用程序所替换，不同于其他手机操作系统固化在系统内部的系统软件，它们更加灵活和个性化。

（2）应用框架层

应用框架层（Application Framework）是 Android 开发的基础，很多核心应用程序是通过这一层来实现其核心功能的。该层简化了组件的重用，开发人员可以直接使用其提供的组件来进行快速的应用程序开发，也可以通过继承而实现个性化的拓展。

- 活动管理器（Activity Manager）：管理各个应用程序生命周期及通常的导航回退功能。
- 窗口管理器（Window Manager）：管理所有的窗口程序。

- 内容提供器（Content Provider）：使得不同应用程序之间存取或者分享数据。
- 视图系统（View System）：构建应用程序的基本组件。
- 通告管理器（Notification Manager）：使得应用程序可以在状态栏中显示自定义的提示信息。
- 包管理器（Package Manager）Android：系统内的程序管理。
- 电话管理器（Telephony Manager）：管理所有的移动设备功能。
- 资源管理器（Resource Manager）：提供应用程序使用的各种非代码资源，如本地化字符串、图片、布局文件、颜色文件等。
- 位置管理器（Location Manager）：提供位置服务。
- XMPP 服务（XMPP Service）：提供 Google Talk 服务。

（3）系统库

系统库（Libraies）是指一些提供底层功能支持的库（主要由 C/C++编写的），是连接应用程序框架层与 Linux 内核层的重要纽带。

系统库各个部分的功能如下。

- Surface Manager：执行多个应用程序时候，负责管理显示与存取操作间的互动，另外也负责 2D 绘图与 3D 绘图进行显示合成。
- 多媒体库（Media Framework）：基于 PacketVideoOpenCore;支持多种常用的音频、视频格式录制和回放，编码格式包括 MPEG4、MP3、H.264、AAC、ARM。
- SQLite：小型的关系型数据库引擎。
- OpenGL|ES：根据 OpenGLES 1.0API 标准实现的 3D 绘图函数库。
- FreeType：提供点阵字与向量字的描绘与显示。
- WebKit：网页浏览器的软件引擎。
- SGL：底层的 2D 图形渲染引擎。
- SSL：在 Andorid 上通信过程中实现握手。
- Libc：从 BSD 继承来的标准 C 系统函数库，专门为基于 embedded linux 的设备定制。

（4）Android 运行时

Android 运行时（Android Runtime）的应用程序是采用 Java 语言编写，但它并不使用 J2ME 来执行程序，而是在 Android 自带的 Android 运行时中执行。Android 运行时分为 Java 核心库和 Dalvik 虚拟机两部分。

Java 核心库：Java 核心库提供了 Java 应用程序接口（Application Programming Interface，API）中的大多数功能，同时也包含了 Android 的一些核心 API，如 android.os、android.net、android.media 等。

Dalvik 虚拟机：Android 程序不同于 J2ME 程序，每个 Android 应用程序都有一个专有的进程，并且不是多个程序运行在一个虚拟机中，而是每个 Android 程序都有一个 Dalvik 虚拟机的实例，并在该实例中执行。Dalvik 虚拟机是一种基于寄存器(register_based)的虚拟机，不是传统的基于栈(stack-based)的虚拟机，进行了内存资源使用优化及支持多个虚拟机的特点。

需要注意的是：不同于 J2ME，Android 程序在虚拟机中执行的并非编译后的字节码，而是通过转换工具 dx 将 Java 字节码转成 dex 格式的中间码。

（5）Linux 内核层

Linux 内核层（Linux Kernel）主要指 Android 寄宿的 Linux 操作系统及相关驱动。通常来说，只有硬件厂商和从事 Android 移植的开发者，才会基于此来进行开发。Android 是基于 Linux2.6 内核，其核心系统服务如安全性、内存管理、进程管理、网路协议以及驱动模型都依赖于 Linux 内核。

任务实现

1. 下载工具包

Android 开发环境搭建之前需要在网上下载以下开发工具包，做好环境配置的准备工作。

（1）JDK 工具包的下载

在进行 Android 开发之前，需要下载安装 Java 的开发环境，因为 Android SDK 的应用层是 Java 语言，所以需要 Java 开发环境。有过 Java 开发经验的读者，对 JDK 的配置应该并不陌生。本书使用的 JDK 是 32 位的 JDK6。

JDK 下载地址是 http://www.java.net/download/jdk6/6u10/promoted/b32/binaries/jdk-6u10-rc2-bin-b32-windows-i586-p-12_sep_2008.exe

（2）下载 ADT 集成开发工具

http://wear.techbrood.com/sdk/index.html 为 Android 在国内的镜像网址。最新的 Android 开发工具和都会在该网址发布，也是 Android 学习者必须了解的网站。目前该网站提供了一个集成各种必备插件的 ADT Bundle 下载，包括 Eclipse、Android SDK 和 ADT 等系列软件包。打开 http://wear.techbrood.com/sdk/index.html，如图 1-3 所示，可下载 Android 开发工具。

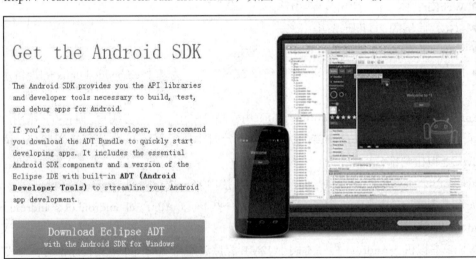

图 1-3　Android 开发工具下载

2. 安装配置 JDK

JDK 开发工具包是 Java 和 Android 应用程序开发的必备工具。从网上下载 JDK 工具包之后进行安装，其安装步骤较简单，基本按照默认设置即可。现有的 Java 开发书籍中对 JDK 的安装都有介绍，此处不再赘述。

安装好 JDK 后，需要进行 JDK 环境变量配置。右键单击"我的电脑"→"属性"→"高级"→"环境变量"→"系统变量"。系统变量设置需要设置 3 个环境变量，其中 JAVA_HOME 的内容为 JDK 的安装路径，ClassPath 指向 Java 的类，Path 指向 Java 的编译、运行等命令工具。具体配置如下，等号左侧的是环境变量名，等号右侧的是环境变量值。

- JAVA_HOME= C:\Program Files\Java\jdk1.7.0_09
- ClassPath=.;%JAVA_HOME%\lib\dt.jar;%JAVA_HOME%\lib\tools.jar;
- Path=%JAVA_HOME%\bin;

注意：其中点"."和分号";"不能缺少，而且必须是英文。

3．安装配置 Eclipse 和 ADT

在 Android 的早期开发中，需要配置 Eclipse、Android SDK 和 ADT。但是现在已经方便很多，可以直接从 Android 的官网下载开发包。该开发包已经将 Android 开发工具集成在一个压缩包中，名为 Android adt-bundle-windows-x86。压缩包中包含 Eclipse+ADT、Android SDK、AVD 和 Android 系统平台。安装配置 Eclipse 和 ADT 步骤如下。

① 解压网上下载的"adt-bundle-windows-x86"压缩包。其中包含已经配置 ADT 的 eclipse3.8 工具包和 Android SDK 文件夹。解压后文件结构如图 1-4 所示。本书使用的是 Android4.2 版本。可以利用 SDK 文件夹下的 SDK Manage 下载需要的 Android 版本。

图 1-4　adt-bundle-windows-x86 文件夹结构

② 启动 ADT，因为 ADT 已经集成在 Eclipse 中，所以启动 ADT 的步骤如下。

a. 打开 eclipse 文件夹，如图 1-5 所示，鼠标左键双击 eclipse.程序。

b. 设置工作空间（workspace）如图 1-6 所示。进入 ADT 后的界面如图 1-7 所示，进入该界面之后就可以进行 Android 应用程序开发了。具体开发过程将在下一个任务中介绍。

图 1-5　eclipse 文件夹结构

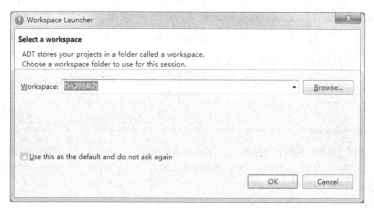

图 1-6　选择 Android 开发工作空间（Workspace）

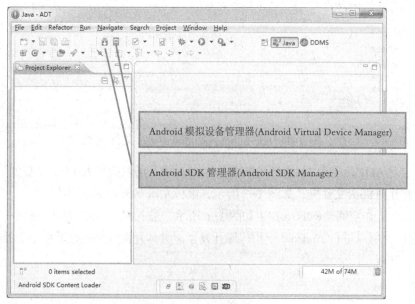

图 1-7　ADT 开发环境

4. 配置 Android SDK

Android SDK 是用于支持 Android 开发的软件工具包，adt-bundle-windows-x86 中已经包含 Android 4.2 的 SDK。一般系统会自动获取 Android SDK 的路径，如果系统未自动获取则需手动配置。为了便于后续的开发工作，这里将讲述如何配置 Android SDK 的路径及如何升级 Android SDK。

（1）配置 Android SDK 路径

如果 ADT 中没有自动配置 Android SDK 路径，则需要手动配置 Android SDK 路径。其步骤为，打开"ADT"→"选择 Windows"→选择"Preferences"选项，如图 1-8 所示。在弹出的图 1-9 所示界面中选中"Android"属性，并在 SDK Location 输入框中选择之前解压缩的 Android SDK 目录→单击 Apply 按钮，再单击 OK 按钮。

提示：SDK Location 为解压 adt-bundle-windows-x86 压缩包后 adt-bundle-windows-x86 中 Android SDK 文件夹的路径。

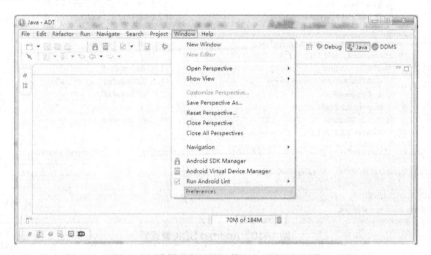

图 1-8　选择 Windows 的 Preferences 项

图 1-9　Android SDK 设置

（2）升级 Android SDK

单击 ADT 集成开发环境主页面的"Android SDK Manager"项，如果网络连接正常可用的话，"Android SDK Manager"窗体中将列出当前已经安装的 Android SDK 软件和其他可用版本的 Android SDK 列表，如图 1-10 所示。

选择要安装的选项，单击"Install3packages"，弹出图 1-11 所示界面，选择"Accept All"。此时就会通过网络从 Google 服务器下载这些软件包。视所选软件包的多少和网速快慢，下载时间可能耗费数小时之久，数据量为 2~5GB。

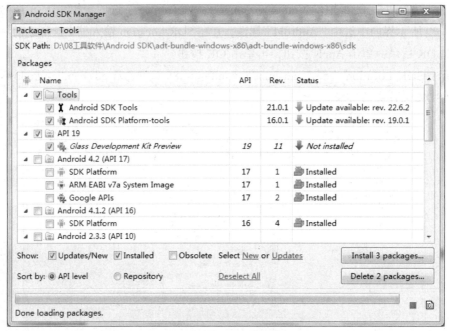

图 1-10　Android SDK 管理器

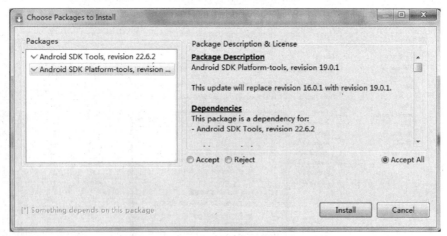

图 1-11　Android SDK 安装选择工具包

5．配置启动 AVD

Android 开发需要配置 AVD 以运行 Android 应用程序。AVD 的配置步骤如下。

① 在打开的 ADT 中选择 按钮，弹出界面如图 1-12 所示，单击 New 按钮开始对 AVD

进行配置，如图 1-13 所示，其配置参数如下。
- AVDName：按照规范命名，此处命名为 android4.2。
- Target：Android 4.2– API Level17。

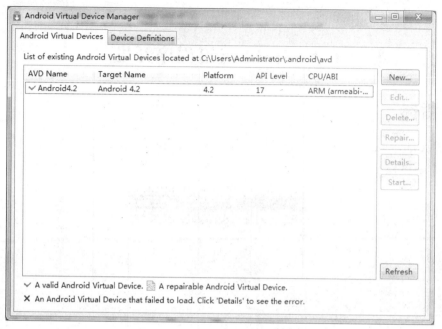

图 1-12　Android AVD 设备管理

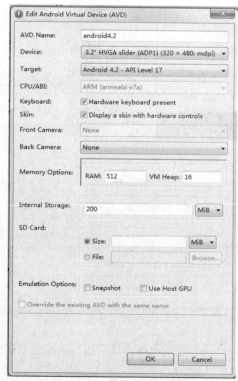

图 1-13　Android AVD 设置图

图 1-14　AVD 运行效果图

② AVD 设置完成之后，启动 AVD，启动步骤如下。

在 Eclipse 中选择 Android Virtual Device Manager 按钮，弹出界面如图 1-12 所示，选择 Android 4.2 项，单击 Start 按钮，弹出虚拟设备的启动界面，如图 1-14 所示，表示 AVD 启动完成。

任务 1.2　第一个 Android 应用程序开发

任务描述

创建 Android 应用程序，界面上显示"明天天气，晴，温度 2~15"，并显示对应图片。其效果如图 1-15 所示。

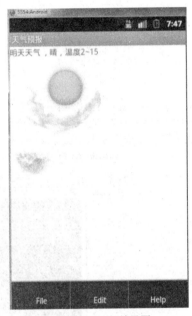

图 1-15　运行效果图

任务目标

① 掌握资源的定义和使用方法。
② 理解 Android 应用程序的框架。
③ 掌握 AndroidManifest 的配置。

任务分析

该任务功能首先要在想到模式下创建一个 Android 应用程序，然后按照功能要求在应用程序中使用字符串与图片资源，具体实现过程如下。

① 在向导模式下创建 Android 应用程序。

② 在代码中定义并使用图片和字符串资源。
③ 配置 AndroidManifest.xml 文件。

知识要点

1. Android 应用程序结构

Android 应用程序的组成结构如图 1-16 所示，在开发应用程序时经常要用到的内容有 src 目录下的 Java 文件、res 目录下的资源文件和 AndroidManifest.xml 文件中的配置信息。

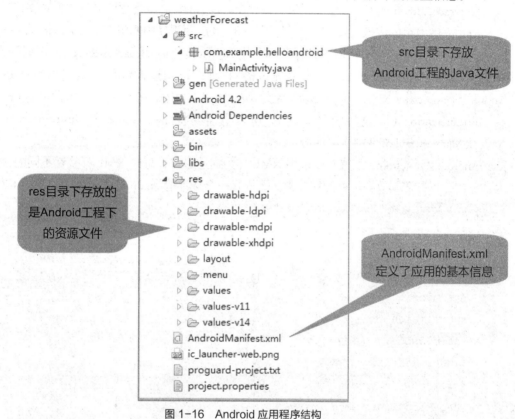

图 1-16　Android 应用程序结构

下面详细介绍每个目录中的内容。

src 目录下存放的是 Android 工程的 Java 文件，在此编写 Java 代码以实现程序要求的功能。

gen 目录下存放的文件是在编译 Android 程序时系统自动生成。在编写程序时不要更改该目录下的文件。

bin 目录下是存放 Android 应用程序通过编译之后生成的文件，如 Android 安装包（Android Package，简称为 apk）就是在此目录之下。

libs 目录下是存放支持 Android 应用程序开发的程序包，比如要进行 JSON 数据解析时需要引入支持 JSON 开发的程序包，引入工具包之后则存放于该目录之下。

res 目录下存放的是 Android 工程的资源文件，包括图片资源、布局资源、字符串资源、菜单资源等。在 Android 应用程序设计过程中，我们经常要用到一些资源文件，如用图片设置界面、用音频设置铃声、用动画显示特效、用字符来提示信息,那么这些图片、音频、动画、字符串就叫做 Android 应用程序中的资源文件。其中 drawable 中是图片资源,layout 中是布局资源,menu 中是菜单资源、values 中是字符串、尺寸、样式等资源。

assets 也是用于存放 Android 工程的资源文件,一般用于保存原生的文件。例如,一个 MP3 文件,Android 程序不能直接访问,必须通过 AseetManager 类以二进制流的形式来读取。

AndroidManifest.xml 文件定义了 Android 应用程序的配置信息。

2. AndroidManifest 配置文件

AndroidManifest.xml 位于整个项目的根目录,是 Android 应用程序必备的配置文件,用于定义应用程序中组件信息。这个文件描述了 Android 应用程序中的组件,如 activities、services、provider 等。一般情况下 Android 应用程序是由 Activity 活动、BroadcastReciver 广播接收器、Service 服务、Content Provider 内容提供器 4 种组件构造而成的,这四类组件必须在 AndroidManifest.xml 文件中声明注册。此外,Android 应用程序在必要时还需指定 permissions 和 instrumentation(安全控制和测试),所以也需要将 permissions 和 instrumentation 在 AndroidManifest.xml 文件中声明注册。

AndroidManifest.xml 配置文件采用 XML 作为描述语言,每个 XML 标签都不同的含义,大部分的配置参数都放在标签的属性中。下面是一个标准的 AndroidManifest.xml 文件样例。

```
<?xml version="1.0" encoding="utf-8"?>
<manifest>
    <!-- 基本配置 -->
    <uses-permission />
    <permission />
    <permission-tree />
    <permission-group />
    <instrumentation />
    <uses-sdk />
    <uses-configuration />
    <uses-feature />
    <supports-screens />
    <compatible-screens />
    <supports-gl-texture />
    <!-- 应用配置 -->
    <application>
     <!-- Activity 配置 -->
     <activity>
       <intent-filter>
         <action />
         <category />
```

```xml
            <data />
        </intent-filter>
        <meta-data />
    </activity>
    <activity-alias>
        <intent-filter> . . . </intent-filter>
        <meta-data />
    </activity-alias>
    <!-- Service 配置 -->
    <service>
        <intent-filter> . . . </intent-filter>
        <meta-data/>
    </service>
    <!-- Receiver 配置 -->
    <receiver>
        <intent-filter> . . . </intent-filter>
        <meta-data />
    </receiver>
    <!-- Provider 配置 -->
    <provider>
        <grant-uri-permission />
        <meta-data />
    </provider>
    <!-- 所需类库配置 -->
    <uses-library />
  </application>
</manifest>
```

下面详细介绍本任务涉及的结点内容。

<manifest>是 AndroidManifest.xml 的根元素，必须包含一个<application>元素并且指定 xlmns:android 和 package 属性。xlmns:android 指定了 Android 的命名空间，默认情况下是 http://schemas.android.com/apk/res/android。Package 是标准的应用包名，以任务中的包名为例，即为"com.example.helloandroid"。此外，sharedUserId 表明数据权限、sharedUserLabel 一个共享的用户名、versionCode 是给设备程序识别版本、versionName 版本显示名称、installLocation 安装参数。

<manifest>标签语法范例如下。

```
<manifest xmlns:android="http://schemas.android.com/apk/res/android"
    package="string"
    android:sharedUserId="string"
```

```
android:sharedUserLabel="string resource"
android:versionCode="integer"
android:versionName="string"
android:installLocation=["auto" | "internalOnly" | "preferExternal"] >
.. ...
/manifest>
```

  \<uses-sdk>用于指定 Android 应用中所需要使用的 SDK 的版本，比如该应用程序必须运行于 Android 2.1 以上版本的系统 SDK 之上，那么就需要指定应用支持最小的 SDK 版本数为 7。每个 SDK 版本都会有指定的整数值与之对应，比如我们最常用的 Android 2.2.x 的版本数是 8。除了可以指定最低版本之外，\<uses-sdk>标签还可以指定最高版本和目标版本，语法范例如下。

```
<uses-sdk android:minSdkVersion="integer"
    android:targetSdkVersion="integer"
    android:maxSdkVersion="integer" />
```

  \<application>元素是\<manifest>的子元素，一个 AndroidManifest.xml 中必须含有一个\<application>标签。这个标签声明了每一个应用程序的组件及其属性(如 icon、label、permission 等)。Android 应用程序的四大组件\<activity>、\<service>、\<provider>、\<receiver> 都在此元素内声明。

  \<activity>元素是\<application>的子元素，\<activity>元素用于定义 Acitivity 信息，应用程序中的 Acitivity 必须在此注册声明后才能使用。

  \<intent-filter>是\<activity>的子元素，该元素包括 action 与 category 两个子元素。action 元素用于定义了一系列常用的动作，其属性 name 值为 android.intent.action.MAIN，表明此 activity 是作为应用程序的入口。category 元素用于指定当前动作（action）被执行的环境，其 name 属性的常见值为 android.intent.category.LAUNCHER，该值决定应用程序是否显示在程序列表里。

  此外，在 AndroidManifest.xml 样例中的\<service>、\<provider>、\<receiver>等元素会在后续的任务中用到，在此不再详细介绍。

3. 资源文件的定义与使用

  Android SDK 中提供了大量的系统资源，如图 1-17 所示，包括布局文件、字符串资源等，都存在于 Android SDK 的 data\res 目录下。

  资源是 Android 应用的重要组成部分，资源主要包括字符串资源、绘画资源、颜色资源、布局资源、尺寸资源等。资源都存放在应用的 res 目录下，大多数资源都将在 R 类中对应一个内部类，定义了其 ID 值。资源的使用一般都有两种方式，即 Java 代码中通过 R.资源类型.ID 形式使用，XML 文件中通过@资源类型/ID 形式使用。

图 1-17 Android 系统资源

（1）字符串资源

字符串资源必须放在 res/values 目录下的 xml 文件中，如图 1-18 所示。使用<string name="">...</string>定义，name 指字符串资源的 key 值。

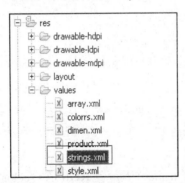

图 1-18 程序框架中字符串资源分布图

```
<?xml version="1.0" encoding="utf-8"?>
<resources>
<string name="hello">大家好，欢迎学习 Android! </string>
<string name="app_name">SDK 中的资源</string>
<string name="weather">明天天气，晴，温度 2~15</string>
</resources>
```

字符串资源的使用：字符串资源中的 key 值是 R.string 类中定义的 int 型的 ID 值。

```
public static final class string {
        public static final int app_name=0x7f040002;
        public static final int weather =0x7f040000;
        public static final int hello=0x7f040001;
}
```

在 Java 代码中使用：R.string.ID 值 text.setText(R.string. weather);

在 XML 中使用：@string/ID 值:@string/ weather

（2）数组资源

在 res/values 下创建表示数组资源的 XML 文件，可以包括字符串数组和整型数组两种，分别使用<string-array>和<integer-array>标签设置。

```
<?xml version="1.0" encoding="utf-8"?>
<resources>
<string-array name="province">
<item>河北省</item>
<item>山西省</item>
<item>辽宁省</item>
</string-array>
<integer-array name="count">
<item>10</item>
<item>20</item>
<item>30</item>
</integer-array>
</resources>
```

数组资源的使用：在 Activity 类中使用 getResources.getStringArray 获得 String 型数组资源。

```
String[] country=this.getResources().getStringArray(R.array.province);
```

在 Activity 类中使用 getResources.getIntArray 方法获得 Int 型数组资源。

```
int[] count=this.getResources().getIntArray(R.array.count);
```

（3）颜色资源

在 XML 文件中使用<color name="">颜色 RGB 值</color>方式保存颜色值，XML 文件按照规范命名，存放在 res\values 目录下。

```
<?xml version="1.0" encoding="utf-8"?>
<resources>
<color name="red">#FF0000</color>
<color name="green">#66FF00</color>
<color name="white">#FFFFFF</color>
</resources>
```

颜色资源的使用：<color name="">中的 name 是 R.color 类中的 ID 值在 Java 代码中使用颜色资源——R.color.ID 值。

```
text.setTextColor(R.color.green);
text.setBackgroundColor(R.color.white);
```

在 XML 中使用颜色资源：@color/ID 值。

```
<TextView android:background="@color/red">
```

（4）尺寸资源

可以在 res\values 下的 XML 文件中使用<dimen name="">浮点数值</dimen>定义尺寸资源。dimen 的值是一系列的浮点数，后面是尺寸单位，常用的单位有：px、in、mm、pt、dp、sp。

```
<?xml version="1.0" encoding="utf-8"?>
<resources>
<dimen name="pxsize">100px</dimen>
<dimen name="insize">10in</dimen>
<dimen name="spsize">100sp</dimen>
</resources>
```

尺寸资源的使用：在 Java 代码中使用 dimen：R.dimen.ID 值，在 XML 中使用 dimen：@dimen/ID 值。

```
<TextView android:layout_height="@dimen/pxsize"
```

（5）样式资源

如果多个组件都需要设置同样的风格，可以在 res\values 下使用 XML 文件存储样式资源，样式使用<style name="">标签指定。

```
<style name="project1style">
<item name="android:gravity">right</item>
<item name="android:background">@color/green</item>
</style>
```

Android 组件可以通过 style 属性指定需要使用的样式资源。

```
<TextView style="@style/project1style"/>
```

（6）类型资源

如果多个组件都需要设置同样的风格，可以在 res\values 下使用 XML 文件存储类型资源，类型使用<style name="">标签指定。

```
<style name="project1style">
<item name="android:gravity">right</item>
<item name="android:background">@color/green</item>
</style>
```

类型资源的使用：Android 组件可以通过 style 属性指定需要使用的类型资源。

```
<TextView style="@style/project1style"/>
```

（7）绘画资源及其使用

Android 应用中会用到很多图像，图像可以放在 res\drawable 下，图片的名称即 ID 值，在 Java 代码中获得 Drawable 对象。

```
Drawable moonpic=this.getResources().getDrawable(R.drawable.moon);
```

在 XML 文件中使用 Drawable 对象。

```
<TextView
android:background="@drawable/moon">
```

（8）布局资源及其使用

Android 应用程序有两种生成组件的方式，即 Java 代码和 XML 文件，所有的 XML 布局文件都存放在 res/layout 下，可以使用 R.layout.ID 值引用布局资源。

```
<?xml version="1.0" encoding="UTF-8"?>
<LinearLayout
android:layout_height="fill_parent"
android:layout_width="fill_parent"
android:orientation="vertical"
mlns:android="http://schemas.android.com/apk/res/android">
<TextView
android:background="@color/red"
android:id="@+id/text"
android:layout_height="@dimen/pxsize"
android:layout_width="fill_parent"
android:text="@string/hello"/>
</LinearLayout>
```

（9）ASSETS 资源及其使用

ASSETS 资源在与 res 目录平级的 assets 目录下，ASSETS 资源不会生成资源 ID，直接使用资源名读取资源文件。

```
try {
InputStream is=this.getAssets().open("LayoutActivity.java");
byte[] buffer=new byte[1024];
int c=is.read(buffer);
String s=new String(buffer,0,c);
ophonetext.setText(s);
} catch (IOException e) {
// TODO Auto-generated catch block
e.printStackTrace();
}
```

总之，配置 Android 应用的开发环境，主要包括 Java SDK、Eclipse、 Android SDK、ADT 插件。几个主要部分运行 Android 应用前，需要配置并启动 AVD 设备模拟器。一个 Android 应用主要包括 Java 源文件（src 目录下）、资源文件（res 目录下）以及配置信息文件 AndroidManifest.xml。

任务实现

1．在向导模式下创建 Android 应用程序

在图 1-7 所示的 Android 开发界面中，选择 "File" → "New" → "Android Application Project"，弹出图 1-19 所示界面。输入应用程序名称 HelloAndroid。在 Mimimum Required SDK 设置 SDK 最小需求版本 API 8：Android 2.2（Frogo），在设置目标 Target SDK 为 API17：Android 4.2(Jelly Bean)，编译的版本 Compile With SDK 为 API17:Android 4.2(Jelly Bean)，主题 Theme：Holoc light with Dark Action Bar。向导模式的操作步骤按照默认步骤依次进行。其执行过程如图 1-20~图 1-23 所示。

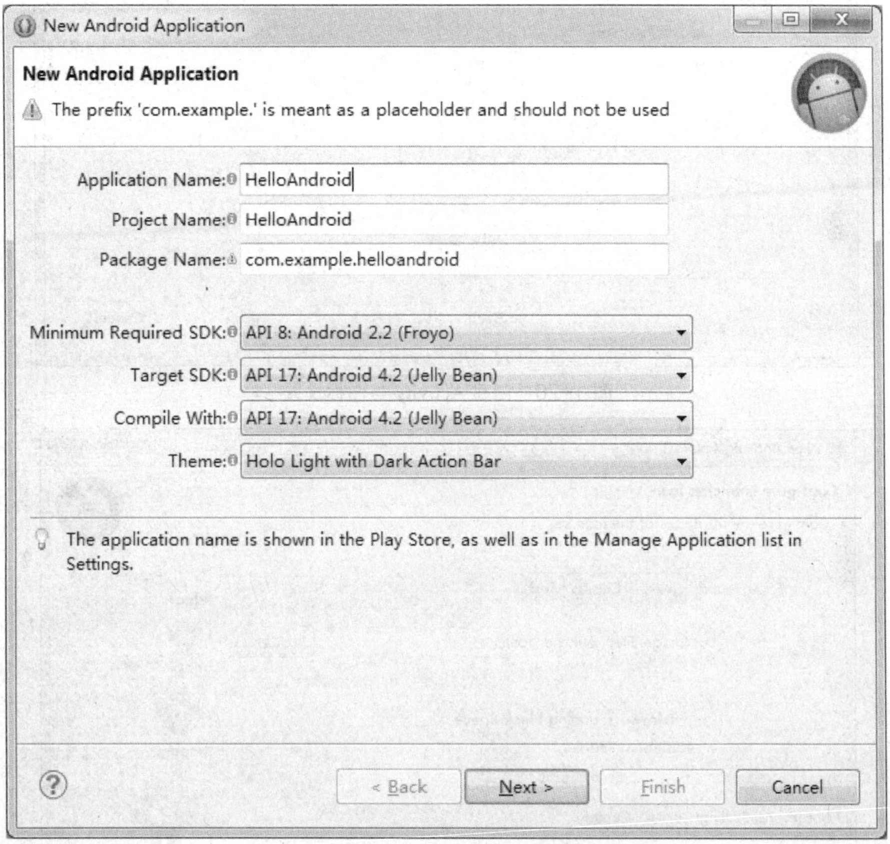

图 1-19　创建 Android 新工程

按照向导模式的步骤依次执行，直到单击 "Finish" 按钮，Android 应用程序基本创建完成。为了实现任务的功能需求要对代码进行相应的设置，下面详细介绍代码设置。

2．在代码中定义并使用图片和字符串资源

将 "明天天气，晴，温度 2~15" 以 weather 为名设置在 strings.xml 资源文件夹当中，并在 Java 代码中通过 R.string.weather 访问。

（1）字符串资源定义与引用

在 res\values\strings 中添加字符串：

图 1-20　创建 Activity 与图标设置

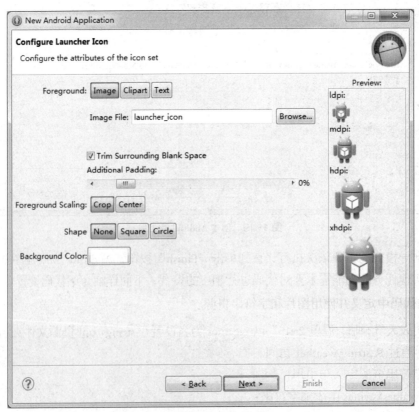

图 1-21　图标属性设置

图 1-22 创建 Activity 选项

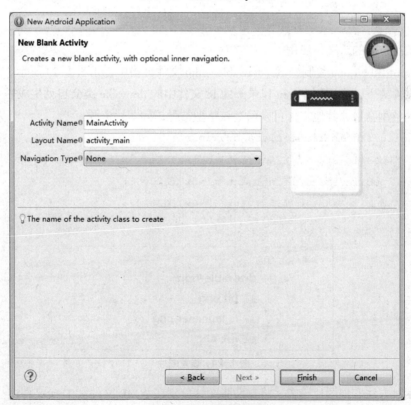

图 1-23 完成工程创建

```xml
<string name="weather">明天天气，晴，温度 5~15</string>
```
字符串资源引用方法有两种，下面介绍两种引用方式。

● 方法一：在 layout 文件夹下的 activity_main.xml 文件中引用。

```xml
<TextView
        android:id="@+id/textView1"
        android:layout_width="wrap_content"
        android:layout_height="wrap_content"
        android:layout_centerHorizontal="true"
        android:layout_centerVertical="true"
        android:text="@string/weather" />
```

● 方法二：在 src 文件夹中 MainActivity.java 文件中设置如下代码。

```java
@Override
    protected void onCreate(Bundle savedInstanceState) {
        super.onCreate(savedInstanceState);
        //setContentView 设置该 Activity 的现实界面
        setContentView(R.layout.activity_main);
        TextView mytext = (TextView)this.findViewById(R.id.textView1);
        //修改文本框中原有的内容
        mytext.setText(R.string.weather);
    }
```

（2）图片资源的定义与使用

将 m1.png 和 bit.png 复制到 res/drawable-hdpi 文件夹当中，如图 1-24 所示。当图片文件复制到该文件夹下，应用程序 gen 目录下的 R 文件中的 drawable 类会自动生成一个名字与图片名相同的整型常量，其他文件可以通过该整型常量访问图片资源。

```java
    public static final class drawable {
        public static final int ic_launcher=0x7f020000;
        public static final int m1=0x7f020001;
        public static final int m2=0x7f020002;
}
```

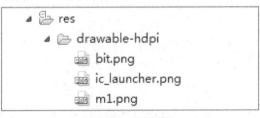

图 1-24 设置图片

利用两种方法实现图片设置，将这两张图片分别在 XML 文件和 Java 文件中引用，XML 为可扩展标记语言(Extensible Markup Language，XML)。

● 方法一：使用在 XML 文件引用的方式。

将 layout 文件夹中的 activity_main.xml 文件中的代码设置如下。

ImageView 控件是一种用于图片显示的控件。

```xml
<ImageView
        android:id="@+id/bitmapView"
        android:layout_width="wrap_content"
        android:layout_height="wrap_content"
        android:layout_centerHorizontal="true"
        android:layout_centerVertical="true"
        android:src="@drawable/m1"
        />
```

● 方法二：使用在 Java 文件引用的方式。

先在 activity_main.xml 中定义 ImageView 控件，然后在 Java 代码中通过获取 R.id 的方式获取该 ImageView，并在 Java 代码中定义图片获取图片，将图片加入到 ImageView 当中。其主要代码如下所示。

```xml
<ImageView
        android:id="@+id/bitmapView1"
        android:layout_width="wrap_content"
        android:layout_height="wrap_content"
        android:layout_centerHorizontal="true"
        android:layout_centerVertical="true"
        />
```

（3）在 MainActivity 中编写代码实现图片加载

在 MainActivity 类中创建 ImageView 对象。

```java
private ImageView myImageView;
```

在 onCreate()方法定义如下内容，实现图片显示。

```java
myImageView = (ImageView)findViewById(R.id.bitmapView1);
Resources r = getResources();
Drawable d = r.getDrawable(R.drawable.bit);
myImageView.setImageDrawable(d);
```

3. 配置 AndroidManifest.xml 文件

Android 的配置文件 AndroidManifest.xml 内容如下。

```xml
<?xml version="1.0" encoding="utf-8"?>
<manifest xmlns:android="http://schemas.android.com/apk/res/android"
package="com.example.helloandroid"
android:versionCode="1"
android:versionName="1.0">
```

```xml
<uses-sdk
    android:minSdkVersion="7"
    android:targetSdkVersion="17"/>
<application
    android:allowBackup="true"
    android:icon="@drawable/ic_launcher"
    android:label="@string/app_name"
    android:theme="@style/AppTheme">
    <activity
    android:name="com.example.helloandroid.MainActivity"
    android:label="@string/app_name">
    <intent-filter>
        <action android:name="android.intent.action.MAIN"/>
        <category android:name="android.intent.category.LAUNCHER"/>
    </intent-filter>
    </activity>
</application>
</manifest>
```

技能训练

设计程序实现图 1-25 所示效果，要求：通过样式资源的使用，完成文字的颜色、大小的设置；第一个文本字符串内容通过在 XML 的方式引用资源，第二个文本内容通过在 Java 代码中调用 string 资源。

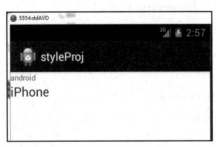

图 1-25 程序运行效果

PART 2 项目二 我爱记单词

项目情境

单词是英语学习的基础,如果能利用手机存储单词,就能随时随地地进行单词的学习。对于英语学习者来说,这样利用碎片时间进行单词记忆能解决纸质资料携带不便的问题,也能提高学习的效率。

本项目是为了实现手机单词管理,使用者能将需要记忆的单词添加到单词库,如果单词信息存在错误可以对单词进行修改和删除。此外,使用者可以按照添加的单词库中的顺序依次查看单词,从而达到对单词的记忆目的。总体来说,该项目需要实现基于 Android 系统的单词管理,包括单词的添加、删除、修改和查看,并实现用户的登录与注册,如图 2-1 所示。通过该项目的任务实现,掌握用户界面设计、数据处理、数据共享、网络互联的基本方法。

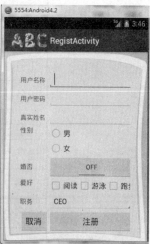

图 2-1 我爱记单词项目运行效果

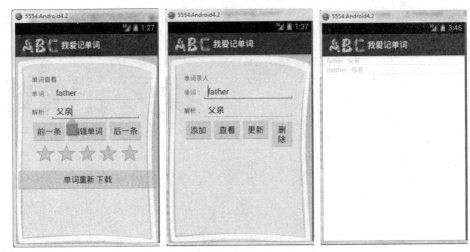

图 2-1 我爱记单词项目运行效果（续）

学习目标

- ☑ 通过用户登录、注册界面的设计，掌握界面设计的基础，包括界面布局和常用控件、事件处理、菜单对话框等相关内容。
- ☑ 通过用户信息存取、单词信息存取，掌握数据存取的几种方法，包括文件存取、SharedPreferences、SQLite、ContentProvider、网络存取。实现单词存取，用户信息的网络传输。

工作任务

任务名称
任务 2.1　用户登录界面设计
任务 2.2　系统退出功能设计
任务 2.3　用户注册界面设计
任务 2.4　用户访问信息存取
任务 2.5　单词存取
任务 2.6　单词共享
任务 2.7　用户信息网络传输
任务 2.8　单词网络下载

任务 2.1　用户登录界面设计

任务描述

"我爱记单词"项目需要记录学习者的相关信息,因此需要实现用户信息的管理,在该任务中实现用户登录界面的设计。其界面设计如图 2-2 所示。

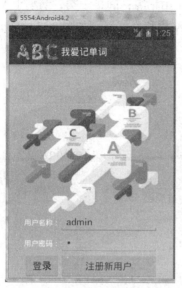

图 2-2　用户登录运行效果

任务目标

① 掌握界面布局和常用控件的使用。
② 掌握资源访问和背景图片设置方法。
③ 掌握事件处理的实现方法。

任务分析

该任务涉及的功能主要有界面设计、资源访问和事件处理,具体实现过程:
① 创建 dict 工程;
② 设计界面布局;
③ 设置背景图片;
④ LoginActivity 中加载控件并实现按钮事件;
⑤ 配置 AndroidManifest.xml 文件。

知识要点

1. View 类简介

View 类主要用于显示界面，Android 应用程序中每个 Activty 都对应一个显示界面，Activity 上展现的是 Android 系统中的可视化组件，Android 中的任何可视化组件都是从 android.view.View 继承。View 类继承于公共父类 Object，View 类有很多子类。View 类的子类分为两种：直接子类，如 TextView、ProgerssBar 等；间接子类，继承于 ViewGroup 的布局类 Layout，如 GridView、ListView。其层次结构和类的简介如图 2-3 所示，该类定义的概要如图 2-4 所示。

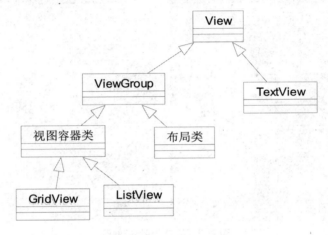

图 2-3 View 类层次结构

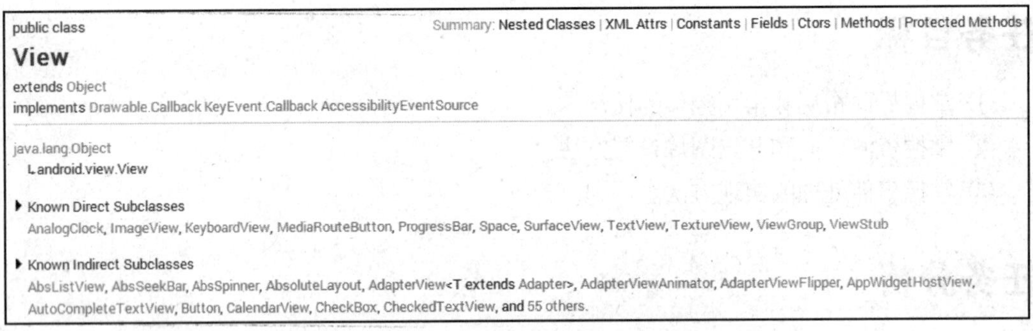

```
public class                         Summary: Nested Classes | XML Attrs | Constants | Fields | Ctors | Methods | Protected Methods
View
extends Object
implements Drawable.Callback KeyEvent.Callback AccessibilityEventSource

java.lang.Object
   ↳ android.view.View

▶ Known Direct Subclasses
   AnalogClock, ImageView, KeyboardView, MediaRouteButton, ProgressBar, Space, SurfaceView, TextView, TextureView, ViewGroup, ViewStub

▶ Known Indirect Subclasses
   AbsListView, AbsSeekBar, AbsSpinner, AbsoluteLayout, AdapterView<T extends Adapter>, AdapterViewAnimator, AdapterViewFlipper, AppWidgetHostView,
   AutoCompleteTextView, Button, CalendarView, CheckBox, CheckedTextView, and 55 others.
```

图 2-4 View 类的结构示意

创建 View 的两种方式：使用 XML 创建 View 和使用 Java 代码创建 View。

- 在 res/layout 使用创建 XML 类型的布局文件，并在 Acitvity 中使用 setContentView()方法将此 XML 文件设置为显示视图。
- 使用代码创建 View，每一个视图组件都是一个 View 类型的对象，可以在代码中使用视图类的构造方法创建 View 对象，调用 View 对象的 setXXX 方法设置其属性。

下面依次介绍这两种方法的实现步骤。

① 使用 XML 文件实现界面设计，步骤如下。

在 res/layout 下创建 activitymian.xml 文件，在该文件中设置布局方式为 LinerLayout（线性

布局），并定义一个按钮控件。

```xml
<?xml version="1.0" encoding="UTF-8"?>
<LinearLayout android:gravity="center_horizontal"
    android:layout_height="fill_parent"
    android:layout_width="fill_parent"
    android:orientation="vertical" xmlns:android="http://schemas.android.com/apk/res/android">
<Button android:id="@+id/btnok"
        android:layout_height="wrap_content"
        android:layout_width="wrap_content"
android:text="确定"/>
</LinearLayout>
```

在 Activity 中将 activitymian.xml 设置为显示界面，使用 setContentView 方法。

```java
public void onCreate(Bundle savedInstanceState) {
super.onCreate(savedInstanceState);
setContentView(R.layout.activitymian);
}
```

在代码中获得 XML 中创建的 View，实际应用中常常需要在代码中获得 XML 中创建的 View 对象（如 TextView、Button、ProgerssBar 等），进一步对其进行操作可以使用 Activity 类的 findViewById( )方法获得 XML 中的 View 对象。

```java
Button button=(Button) this.findViewById(R.id. btnok);
button.setText("按钮");
```

②使用 Java 代码创建界面，步骤如下。

下面代码创建一个 TextView，并将 TextView 设置为显示界面。

```java
public void onCreate(Bundle savedInstanceState) {
    super.onCreate(savedInstanceState);
    TextView tv=new TextView(this);
    tv.setText("您好！Android!");
    setContentView(tv);
}
```

2. 布局管理器

Android 系统中提供了几种常用的布局方式：线性布局、层布局、相对布局、表格布局、坐标布局。每种布局方式都对应 API 中一个类。

LinearLayout：线性布局。

FrameLayout：层叠布局。

RelativeLayout：相对布局。

TableLayout：表格布局。

AbsoluteLayout：坐标布局。

LinearLayout 为线性布局，将组件按照属性设置的水平或垂直方向按顺序排列。设置布局显示方式的属性有：方向 orientation，对齐方式 gravity。orientation 表示控件是水平还是垂直放置，gravity 表示是居中还是居左对齐。在 Layout 中每个控件的 layout_height、layout_width 和 android:orientation 属性是必须定义的，表示控件的高度和宽度属性设置。

```
<LinearLayout
    android:gravity="center_horizontal"
    android:layout_height="fill_parent"
    android:layout_width="fill_parent"
    android:orientation="vertical"
    mlns:android="http://schemas.android.com/apk/res/android">
</LinearLayout>
```

FrameLayout 称为层布局，将组件显示在屏幕的左上角，后面的组件覆盖前面的组件。

```
<FrameLayout
    android:gravity="center_horizontal"
    android:layout_height="fill_parent"
    android:layout_width="fill_parent"
    android:orientation="vertical"
    mlns:android="http://schemas.android.com/apk/res/android">
</FrameLayout>
```

RelativeLayout 是相对布局，相对布局指的是某个组件的位置是相对于其他组件的位置。

```
<RelativeLayout
    android:gravity="center_horizontal"
    android:layout_height="fill_parent"
    android:layout_width="fill_parent"
    xmlns:android="http://schemas.android.com/apk/res/android">
<Button android:id="@+id/button1"
    android:gravity="center_vertical"
    android:layout_height="wrap_content"
    android:layout_width="wrap_content"
    android:text="按钮 1"/>
</RelativeLayout>
```

TableLayout 是表格布局，使用 TableRow 指定一行，每个组件表示一列。

```
<TableLayout
    android:layout_height="fill_parent"
    android:layout_width="fill_parent"
    mlns:android="http://schemas.android.com/apk/res/android">
<TableRow>
```

```xml
<Button android:id="@+id/button1"
    android:layout_height="wrap_content"
     android:layout_width="wrap_content" android:text="按钮1"/>
<Button android:id="@+id/button2"
    android:layout_height="wrap_content"
    android:layout_width="wrap_content" android:text="按钮2"/>
</TableRow>
</TableLayout>
```

AbsoluteLayout 是绝对布局管理器，指的是指定组件的左上角绝对坐标来指定组件的布局，通常需要设置横纵坐标 layout_x 和 layout_y。

```xml
<AbsoluteLayout
    android:gravity="center_horizontal"
    android:layout_height="fill_parent"
    android:layout_width="fill_parent"
    android:orientation="vertical"
    mlns:android="http://schemas.android.com/apk/res/android">
<Button android:id="@+id/button1"
    android:layout_height="wrap_content"
    android:layout_width="wrap_content"
    android:layout_x="0px"
    android:layout_y="0px"
    android:text="按钮1"/>
</AbsoluteLayout>
```

3．文本相关组件

显示文本的组件 TextView，如邮件正文或应用程序标签等。API 中对应 android.widget.TextView 类。

```xml
<TextView
    android:background="#FF00FF"
    android:layout_gravity="center"
    android:layout_height="wrap_content"
    android:layout_margin="60dp"
    android:layout_width="fill_parent"
    android:padding="40dp"
    android:text="@string/hello"
    android:textColor="#0000FF"/>
```

TextView 的重要属性如下。

文本的颜色和背景：

```
android:textColor="#0000FF"
android:background="#0000FF"
```

TextView 的对齐方式：

```
android:layout_gravity="center"
```

字体大小：

```
android:textSize="20px"
```

单行显示：

```
android:singleLine="true"
```

编辑文本的组件 EditText。EditText 是 TextView 类的子类，具有 TextView 所有属性。EditText 可以编辑文本，也可以指定输入文本的类型，通过 3 种属性可以指定：android:digits、android:inputType、android:numeric。

```
<EditText android:background="#FFFFFF"
    android:digits="0123456789"
    android:layout_height="wrap_content"
    android:layout_margin="10dp"
    android:layout_width="200dp"
    android:textColor="#000000"/>
```

自动完成输入内容的组件 AutoCompleteTextView。AutoCompleteTextView 是 EditText 的子类，当用户键入文本时，能提示输入建议。

4．普通按钮

Button 类继承了 TextView 类。在布局文件中，使用属性指定 Button 的属性，如 android：text 指定 Button 显示的文本。

```
<Button android:id="@+id/button1"
    android:layout_height="wrap_content"
    android:layout_width="wrap_content"
    android:text="普通按钮"/>
```

5．事件处理

事件处理的实现方法有 3 种，分别是类本身实现事件处理接口和方法、匿名内部类实现事件处理的方法、一般内部类实现事件处理接口和方法。事件处理的实现有 3 个关键步骤：实现事件处理接口 implementsOnClickListener，实现事件处理的方法 publicvoidonClick(View arg0)和设置监听方法 setOnClickListener()。下面依次介绍事件处理的 3 种实现方法。

① 利用类实现鼠标单击事件接口和方法。在类的定义中利用 implementsOnClickListener 语句实现事件单击事件接口，在类中实现接口的方法 public voidonClick(View arg0)，方法中利用 switch 语句匹配组件产生的事件处理。在按钮事件监听方法 setOnClickListener(this)实现监听。

② 匿名内部类实现鼠标单击事件处理。在按钮事件监听方法 setOnClickListener()中直接利用 newOnClickListener()创建事件处理对象并实现 public voidonClick(View arg0)方法。

③ 一般内部类实现事件处理。创建一般内部类 actLiner，该类定义中利用 implementsOnClickListener 实现接口，在类中实现接口的方法 public voidonClick(View arg0)。在按钮事件监听方法 setOnClickListener(new actLiner())实现监听。

事件处理方法一：利用类。

```
public class LoginActivity extends Activity implements OnClickListener {
    private Button btnok,btnReg;
    private EditText username,password;
    @Override
    protected void onCreate(Bundle savedInstanceState) {
        super.onCreate(savedInstanceState);
        setContentView(R.layout.login);
        btnok=(Button) findViewById(R.id.loginButton);
        btnReg=(Button) findViewById(R.id. regButton);
        username=(EditText) findViewById(R.id.userEditText);
        password=(EditText) findViewById(R.id.pwdEditText);
        //添加事件处理
        btnok.setOnClickListener(this);
    }
    @Override
    public void onClick(View arg0) {
        // TODO Auto-generated method stub
        switch(arg0.getId())
        {
        case R.id.loginButton:
            Toast.makeText(getApplicationContext(), "恭喜你登录成功", 2000).show();
        }
    }
}
```

事件处理方法二：利用匿名内部类。

```
public class LoginActivity extends Activity {
    private Button btnok,btncancle;
    private EditText username,password;
    @Override
    protected void onCreate(Bundle savedInstanceState) {
        super.onCreate(savedInstanceState);
        setContentView(R.layout.login);
        btnok=(Button) findViewById(R.id.loginButton);
```

```java
        btnReg =(Button) findViewById(R.id. regButton);
        username=(EditText) findViewById(R.id.userEditText);
        password=(EditText) findViewById(R.id.pwdEditText);
        //添加事件处理
        btnok.setOnClickListener(new OnClickListener() {
            @Override
            public void onClick(View v) {
                    Toast.makeText(getApplicationContext(),"恭喜你登录成功",2000).show();
            }
        });
    }
}
```

事件处理方法三：利用一般内部类。

```java
public class LoginActivity extends Activity {
    private Button btnok,btncancle;
    private EditText username,password;
    @Override
    protected void onCreate(Bundle savedInstanceState) {
        super.onCreate(savedInstanceState);
        setContentView(R.layout.login);
        btnok=(Button) findViewById(R.id.loginButton);
        btnReg =(Button) findViewById(R.id. regButton);
        username=(EditText) findViewById(R.id.userEditText);
        password=(EditText) findViewById(R.id.pwdEditText);
        //添加事件处理
        btnok.setOnClickListener(new actLiner());
    }
    @Override
    public boolean onCreateOptionsMenu(Menu menu) {
        // Inflate the menu; this adds items to the action bar if it is present.
        getMenuInflater().inflate(R.menu.activity_main, menu);
        return true;
    }
    class actLiner implements OnClickListener {
        @Override
        public void onClick(View arg0) {
            // TODO Auto-generated method stub
```

```
                    Toast.makeText(getApplicationContext(), "恭喜你登录成功",
2000).show();
                }
            }
        }
}
```

任务实现

1. 创建 dict 工程

启动 ADT，在 ADT 中创建工程 dict，设置工程的包名、最小兼容 SDK、目标 SDK 等信息，创建过程如图 2-5~图 2-9 所示。

设置参数如下。

- Application Name（应用名称）：ABC 我爱记单词。
- Project Name（工程名）：dict。
- Package Name（包名）：com.siit.dict。
- Minimum Required SDK（最小兼容 SDK）：API 7：Android2.1（Eclair）。
- Target SDK（目标 SDK）：API17：Android 4.2（Jelly Bean）。
- Compile With（编译版本）：API17：Android 4.2（Jelly Bean）。
- Theme（主题样式）:Holo Light with Dark Action Bar。

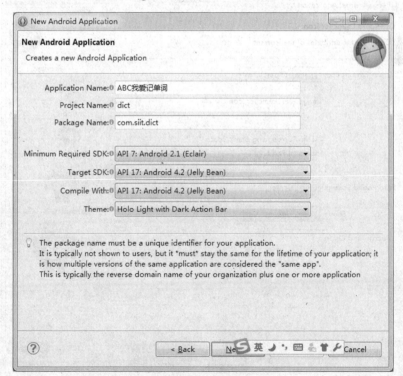

图 2-5　创建工程属性的设置

图 2-6　选中 CreateActivity 和 Create custom launcher icon

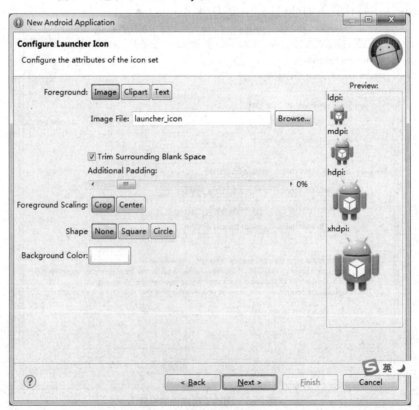

图 2-7　选择图标（icon）

图 2-8 选中创建 BlackActivity

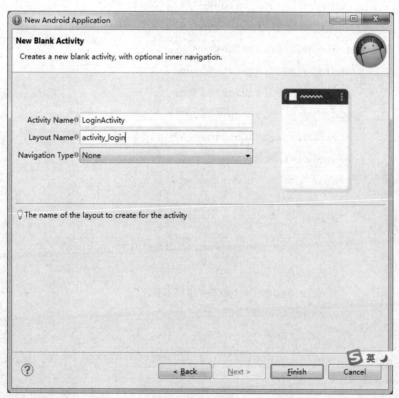

图 2-9 设置 Activity 名称和 Layout 名称

2. 设计界面布局

在 res\layout 文件夹中设计 activity_login.xml 文件,并将控件按照图 2-2 所示的方式设计。在 activity_login.xml 文件中设计两个 TextView 用于显示"用户名称"和"用户密码",设计两个 EidtText 用于输入"用户名称"和"用户密码",设计两个按钮分别为"登录"和"注册新用户"。

```xml
<RelativeLayout xmlns:android="http://schemas.android.com/apk/res/android"
    xmlns:tools="http://schemas.android.com/tools"
    android:layout_width="match_parent"
    android:layout_height="match_parent"
    tools:context=".MainActivity" >
    <LinearLayout
        android:layout_width="fill_parent"
        android:layout_height="fill_parent"
        android:orientation="vertical" >
        <TableLayout
            android:id="@+id/userlogin"
            android:layout_width="fill_parent"
            android:layout_height="fill_parent"
            android:background="@drawable/background"
            android:gravity="bottom"
            android:paddingLeft="30dp"
            android:paddingRight="30dp"
            android:paddingTop="20dp"
            android:stretchColumns="1"
            android:visibility="visible" >
            <TableRow>
                <TextView
                    android:id="@+id/TextView"
                    android:layout_width="wrap_content"
                    android:layout_height="wrap_content"
                    android:text="用户名称: ">
                </TextView>
                <EditText
                    android:id="@+id/userEditText"
```

```xml
                android:layout_width="fill_parent"
                android:layout_height="wrap_content"
                android:text="admin" >
            </EditText>
        </TableRow>
        <TableRow>
            <TextView
                android:id="@+id/TextView"
                android:layout_width="wrap_content"
                android:layout_height="wrap_content"
                android:text="用户密码: " >
            </TextView>
            <EditText
                android:id="@+id/pwdEditText"
                android:layout_width="fill_parent"
                android:layout_height="wrap_content"
                android:password="true"
                android:text="1" >
            </EditText>
        /TableRow>
        <TableRow>
            <Button
                android:id="@+id/loginButton"
                android:layout_width="wrap_content"
                android:layout_height="wrap_content"
                android:text="登录" />
            <Button
                android:id="@+id/regButton"
                android:layout_width="wrap_content"
                android:layout_height="wrap_content"
                android:text="注册新用户" />
        </TableRow>
    </TableLayout>
    </LinearLayout>
</RelativeLayout>
```

3. 设置背景图片

图 2-10 所示为图片属性设置代码：android:background="*@drawable/background*"。

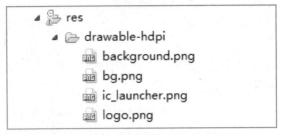

图 2-10　图片资源示意

4. LoginActivity 中加载控件并实现按钮事件

首先定义 btnok、btnReg 两个按钮，并定义 username、password 两个文本编辑框。

在 LoginActivity 的 onCreate 方法中使用 findViewById()方法依次加载按钮和文本控件，并使用 setOnClickListener()方法实现按钮单击事件的监听，其代码如下所示。

```java
public class LoginActivity extends Activity implements OnClickListener {
    private Button btnok,btnReg;
    private EditText username,password;
    @Override
    protected void onCreate(Bundle savedInstanceState) {
        super.onCreate(savedInstanceState);
        setContentView(R.layout.activity_login);
        btnok=(Button) findViewById(R.id.loginButton);
        btnReg=(Button) findViewById(R.id.regButton);
        username=(EditText) findViewById(R.id.userEditText);
        password=(EditText) findViewById(R.id.pwdEditText);
        //添加事件处理
        btnok.setOnClickListener(this);
    }
    @Override
    public void onClick(View arg0) {
        // TODO Auto-generated method stub
        switch(arg0.getId())
        {
        case R.id.loginButton:
            Toast.makeText(getApplicationContext(), "恭喜你登录成功", 2000).show();
```

```
            break;
        }
    }
}
```

5. 配置 AndroidManifest.xml 文件

AndroidManifest.xml 文件在创建工程时系统会自动生成。如果要更改信息,可以设置相关代码,此处修改应用程序的显示图标,将 android:icon 设置为"@drawable/logo。

```xml
<?xml version="1.0" encoding="utf-8"?>
<manifest xmlns:android="http://schemas.android.com/apk/res/android"
    package="com.siit.dict"
    android:versionCode="1"
    android:versionName="1.0" >
    <uses-sdk
        android:minSdkVersion="7"
        android:targetSdkVersion="17" />
    <application
        android:allowBackup="true"
        android:icon="@drawable/logo"
        android:label="@string/app_name"
        android:theme="@style/AppTheme" >
        <activity
            android:name="com.siit.dict.LoginActivity"
            android:label="@string/app_name" >
            <intent-filter>
                <action android:name="android.intent.action.MAIN" />
                <category android:name="android.intent.category.LAUNCHER" />
            </intent-filter>
        </activity>
    </application>
</manifest>
```

技能训练

创建一个 Android 应用程序,该程序可以输入电话号码,单击一个发送按钮,将输入的电话号码显示到当前的界面,程序效果如图 2-11 所示。

图 2-11 程序运行效果

任务 2.2 系统退出功能设计

任务描述

在任务 2.1 的基础上，实现系统退出和系统介绍的功能。其执行效果如图 2-12 所示。单击"About"选项后弹出对话框显示"这是一个用于单词记忆和管理的小程序！"，单击"Exit"选项后显示退出选择框。

图 2-12 程序运行效果

任务目标

① 掌握常见布局的使用。
② 掌握菜单的定义和创建方法。
③ 掌握对话框的定义和使用方法。

任务分析

该任务中用到的资源有字符串、图片、菜单、对话框等内容访问。实现过程中要增加菜单功能，分别实现上下文菜单和选项菜单，并在菜单选择项的单击事件弹出对话框。具体实现过程如下。

① 设置菜单选项。
② 实现"About"系统介绍的方法。
③ 实现"Exit"退出游戏的方法。
④ 在菜单事件中实现"About"与"Exit"功能。

知识要点

1. 对话框

AlertDialog 类继承了 Dialog 类，是其他对话框类的父类。AlertDialog 类有一个重要的内嵌类 Builder。DatePickerDialog、TimePickerDialog、ProgerssDialog 是 AlertDialog 类的子类，如图 2-13 所示。

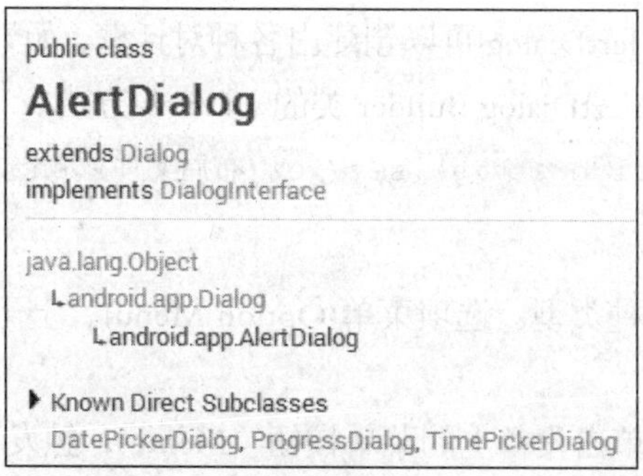

```
public class
AlertDialog
extends Dialog
implements DialogInterface

java.lang.Object
 └ android.app.Dialog
    └ android.app.AlertDialog

▶ Known Direct Subclasses
DatePickerDialog, ProgressDialog, TimePickerDialog
```

图 2-13　Android 中对话框类

以下是对话框上的 3 种按钮。
（1）取消按钮

```
setNegativeButton (CharSequence text, DialogInterface.OnClickListener listener)
setNegativeButton (int textId, DialogInterface.OnClickListener listener)
```

（2）确认按钮

```
setPositiveButton (CharSequence text, DialogInterface.OnClickListener listener)
setPositiveButton (int textId, DialogInterface.OnClickListener listener)
```

（3）覆盖按钮

```
setNeutralButton (int textId, DialogInterface.OnClickListener listener)
setNeutralButton (CharSequence text, DialogInterface.OnClickListener listener)
```

对话框的事件处理，使用 DialogInterface 可提供一系列的内嵌类，监听对话框事件。

```
DialogInterface.OnCancelListener
DialogInterface.OnClickListener
DialogInterface.OnDismissListener
DialogInterface.OnKeyListener
DialogInterface.OnMultiChoiceClickListener
DialogInterface.OnShowListener
```

ProgressDialog 类，表示进度条对话框，ProgressDialog 类结构如图 2-14 所示。ProgressDialog 类的主要方法有：setIcon、setTitle、setMessage、setButton。

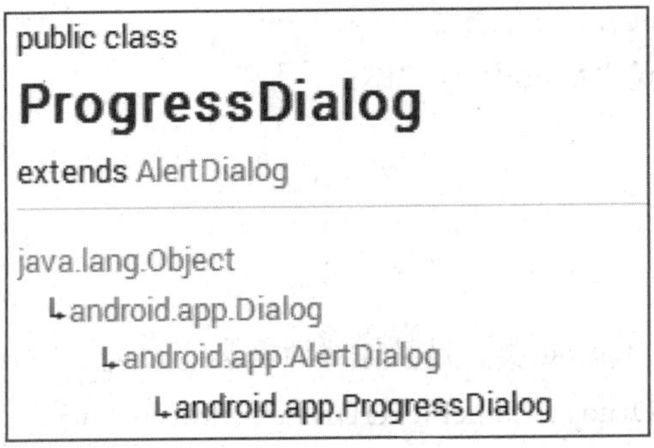

```
public class
ProgressDialog
extends AlertDialog

java.lang.Object
   └ android.app.Dialog
      └ android.app.AlertDialog
         └ android.app.ProgressDialog
```

图 2-14 Android 中 ProgressDialog 类

自定义对话框，使用 AlertDialog 可以创建出各种对话框，如果要完全定制自己的对话框，可以自定义对话框，使用 AlertDialog.Builder 类的 setView 方法。

```
new AlertDialog.Builder(this).setView(布局文件).show()
```

2．菜单

Android 的菜单分为 3 种类型：选项菜单(Option Menu)、上下文菜单(Context Menu)、子菜单(Sub Menu)。

① 选项菜单。当用户单击设备上的菜单按钮（Menu），触发事件弹出的菜单就是选项菜单。选项菜单最多只有 6 个，超过 6 个时，第六项就会自动显示为"更多选项"。

创建方法如下。

- 覆盖 Activity 的 onCreateOptionsMenu(Menu menu)方法，当第一次打开菜单时调用。
- 用 Menu 的 add()方法添加菜单项(MenuItem)，可以调用 MenuItem 的 setIcon()方法为菜单项设置图标。
- 当菜单项(MenuItem)被选中时，覆盖 Acitivy 的 onOptionsMenuSelected()方法响应事件。

② 上下文菜单。当用户长按 Activity 页面时，弹出的菜单称为上下文菜单。

创建方法如下。

- 覆盖 Activity 的 onCreateContextMenu()方法，调用 Menu 的 add 方法添加菜单项

MenuItem。
- 覆盖 onContextItemSelected()方法，响应菜单单击事件。调用 registerForContextMenu()方法，为视图注册上下文菜单。

③ 子菜单。子菜单就是将相同功能的分组进行多级显示的一种菜单，比如，Windows 的"文件"菜单中就有"新建""打开""关闭"等子菜单。在选项菜单和上下文菜单都可以添加子菜单，但子菜单不能嵌套子菜单，这就意味着在 Android 系统中，菜单只有两层。设计时需要注意的是，子菜单不支持 icon。

创建方法如下。
- 覆盖 Activity 的 onCreateOptionsMenu()方法，调用 Menu 的 addSubMenu()方法添加子菜单项。
- 调用 SubMenu 的 add()方法，添加子菜单项。
- 覆盖 onCreateItemSelected()方法，响应菜单单击事件。

3．通知

Toast 是一段显示给用户的小文本，不需要用户响应，在规定时间内自动消失，API 中提供了 Toast 类，创建 Toast 对象。

Notification 是显示在屏幕上方状态栏的信息，Notification 需要使用 NotificationManager 来管理。

任务实现

1．设置菜单选项

在 res\menu\activity_login.xml 文件中定义"about"和"exit"两个菜单项，内容如下。

```
<menu xmlns:android="http://schemas.android.com/apk/res/android" >
    <item
        android:id="@+id/about"
        android:title="About"/>
    <item
        android:id="@+id/exit"
        android:title="Exit"/>
</menu>
```

2．实现"About"系统介绍的方法

创建 aboutAlert(String msg)方法，该方法是单个按钮的对话框显示。

```
// 显示对话框
private void aboutAlert(String msg){
    AlertDialog.Builder builder = new AlertDialog.Builder(this);
    builder.setMessage(msg)
        .setCancelable(false)
```

```
            .setPositiveButton("确定", new DialogInterface.OnClickListener() {
             public void onClick(DialogInterface dialog, int id) {
                }
            });
            AlertDialog alert = builder.create();
            alert.show();
}
```

3. 实现"Exit"退出游戏的方法

创建 exitAlert (String msg)方法，实现退出游戏对话框显示。

```
// 显示对话框
private void exitAlert(String msg){
    AlertDialog.Builder builder = new AlertDialog.Builder(this);
    builder.setMessage(msg)
     .setCancelable(false)
            .setPositiveButton("确定", new DialogInterface.OnClickListener() {
        public void onClick(DialogInterface dialog, int id) {
            finish();
            }
          }).setNegativeButton("取消", new DialogInterface.OnClickListener() {
        public void onClick(DialogInterface dialog, int id) {
        return;
        }
    });
    AlertDialog alert = builder.create();
    alert.show();
}
```

4. 在菜单事件中实现"About"与"Exit"功能

当单击 about 和 exit 菜单项时，调用 aboutAlert(String msg)与 exitAlert(String msg)方法实现"About"关于游戏与"Exit"退出游戏的功能。

```
    @Override
    public boolean onCreateOptionsMenu(Menu menu) {
        // Inflate the menu; this adds items to the action bar if it is present.
        getMenuInflater().inflate(R.menu.activity_login, menu);
        return true;
    }
    @Override
    public boolean onOptionsItemSelected(MenuItem item) {
```

```
        switch (item.getItemId()) {
            case R.id.about:
                aboutAlert("这是一个用于单词记忆和管理的小程序！");
                break;
            case R.id.exit:
                exitAlert("真的要退出吗？");
                break;
        }
        return true;
    }
}
```

技能训练

创建 Android 应用程序，程序运行效果如图 2-15 所示。

程序功能要求：

① 该应用程序具有"设置""视频""信息"3 个菜单；

② 单击"信息"菜单，弹出"信息"的子菜单；

③ 单击"删除"子菜单，弹出对话框，询问是否删除收件箱中所有信息。

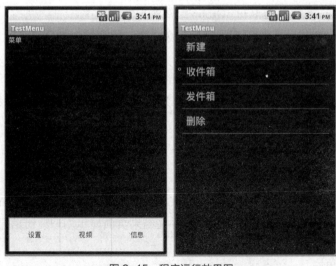

图 2-15 程序运行效果图

任务 2.3 用户注册界面设计

任务描述

注册界面的设计，界面效果如图 2-16 所示，通过 TextView、EditText、RadioGroup、

RadioButton、ToggleButton、CheckBox、Spinner、Button 等控件的使用实现用户注册。

图 2-16 程序运行效果

任务目标

① 会使用 TextView 和 EditText 实现用户名和密码的设置。

② 会使用 RadioButton 与 RadioGroup 组件实现性别设置。

③ 会使用 ToggleButton 按钮实现婚姻状况显示，内容项为"OFF"和"ON"，默认项为"OFF"。

④ 会使用 CheckBox 按钮实现爱好项，选择项为：阅读、游泳。

⑤ 会使用 Spinner 按钮实现职务项，内容为：CEO、CFO、PM，使用数组适配器。

⑥ 会使用 ListView 实现注册内容显示。

任务分析

本任务中使用 TextView、EditText、RadioGroup、RadioButton、ToggleButton、CheckBox、Spinner、Button 等控件实现用户注册页面，并在结果页面显示用户注册信息。具体实现过程如下。

① 创建注册界面。

② 实现注册功能。

③ 实现登录页面向注册页面的跳转。

④ 创建结果界面。

⑤ 创建结果页面的 Activity 文件。

⑥ 配置 AndroidManifest.xml 文件。

知识要点

1. Activity 简介

Activity 是 Android 应用程序的重要组成部分，是 Android 应用程序的入口，也是用户和应用程序之间进行交互的接口。每个 Android 应用程序包含很多 Activity，但只显示在栈顶的 Activity，每个 Activity 中都可以放很多控件，可以把 Activity 看作控件的容器。

创建 Activity 基本步骤如下。

① 自定义创建的 Activity 必须继承自 Activity 类，并导入 android.app.Activity 包。

② 重写 onCreate() 方法，Activity 第一次运行时 Android 操作系统就会自动调用 onCreate 方法。

③ 为 Activity 添加必要的控件，定义控件，并使用 findViewById() 加载 xml 中的控件，并根据功能需求实现相应的代码。

④ 在 AndroidManifest.xml 文件中注册 Activity。每个 Activity 都需要在 AndroidManifest.xml 中注册才能使用。Android 应用程序创建时默认会将当前 Activity 设置为启动 Acitivity，比如本例中的 LoginActivity。当建立新的 Activity 时，将新建的 Activity 在 AndroidManifest.xml 中进行注册，如本例中的 RegistActivity。

```xml
<activity
    android:name="com.siit.dict.LoginActivity"
    android:label="@string/app_name" >
    <intent-filter>
      <action android:name="android.intent.action.MAIN" />
      <category android:name="android.intent.category.LAUNCHER" />
    </intent-filter>
</activity>
<activity android:name="com.siit.dict.RegistActivity"></activity>
<activity android:name="com.siit.dict.ResultActivity"/>
</activity>
```

2. Intent 简介

Intent 是 Android 应用的各项组件之间数据通信的桥梁。当一个 Activity 需要启动另一个 Activity 时，需要通过 Intent 表达意图：需要启动那个 Intent。此外 Android 应用程序通过 Intent 启动 3 个系统组件：Activity、Service、BroadCastReceiver。

Intent 封装 Android 应用程序需要启动某个组件的意图，不仅如此，Intent 还是英语程序之间通信的重要媒介，正如前面程序看到的，两个 Acitivity 可以把需要交换的数据封装成 Bundle 对象，然后使用 Intent 来携带 Bundle 对象以实现两个 Activity 之间的数据传递。

表 2-1 显示使用 Intent 启动 3 种系统组件的方法。

表 2-1 使用 Intent 启动不同组件的方法

组件	启动方法
Activity	starActivity()
	starActivityForResult()
Service	startService()
	bindSerivce()
broadcastReceive	sendBroadcast()
	sendOrderBroadcast()
	sendStickBroadcast()

Activity 方面。Intent 对象通过 Context.starActivity(Intent intent)或 Activity.starActivityForResult(Intent intent, int requestCode())启动一个新的 activity。Activity 的跳转，通过 Extras 传值，当传递的数据较多时通常使用 Bundle 对象实现。

Service 方面。Context.startService()方法，启动新服务或者向正在运行的服务提供新命令。Intent 对象传递到 Context.bindSerivce()中将建立一个服务，建立组件间的联系。

broadcastReceive 方面。Intent 对象通过 Context.sendBroadcast()等发送广播，并将广播内容传递给注册了该广播的广播接收器。

本任务实现过程中演示了如何使用 Intent 实现 Activity 之间的跳转。下面介绍 Intent 对象各个属性的作用。

Intent 负责对应用中一次操作的动作、动作涉及数据、附加数据进行描述。Android 则根据此 Intent 的描述，负责找到对应的组件，将 Intent 传递给调用的组件，并完成组件的调用。Intent 对象抽象地描述了要执行的操作，其描述的基本内容可以分为 Component（组件名称）、Extra（附加信息）、Action（动作）、Category（类别）、Data（数据）、Type（类型）和 Flag（标志位）6 部分，其中 Component 用于明确启动的目标组件，而 Extra 用于"携带"需要传递的数据。

Component 指定 Intent 目标组件的名称。组件名称是一个 ComponentName 对象，这种对象名称是目标组件类名和目标组件所在应用程序的包名的组合。

Component 组件设置方法的简化形式：

```
Intent intent=new(ComponentAttr.this,SecondActivity.class);
```

Extras 是使用 Intent 连接不同组件时在 Intent 中附加额外的信息，以将数据传递给目标 Activity。当传递的数据量较多时，通常以 Bundle 的形式定义再存入 Extras 当中，如本例中的注册页面向结果页面的跳转代码。

```
Bundle b=new Bundle();
b.putString("username", "用户名称"+username.getText().toString());
b.putString("password", "用户密码"+password.getText().toString());
Intent intent=new Intent(RegisterActivity.this, ResultActivity.class);
intent.putExtra("data", b);
startActivity(intent);
```

Action 描述 Intent 所触发动作名字的字符串。例如："android.intent.action.MAIN"表示程序的主入口，不会接收数据，结束后也不返回数据。

Category 用于为 Actiont 增加额外的附加类别信息。Android 在 Intent 类中定义了一组静态常量便是 Intent 不同的类别，如："android.intent.category.LAUNCHER"表示目标 Activity 是应用程序中最优先被执行的 Activity。

Category 与 Action 属性的应用参见本任务程序中的 AndroidMainfest.xml 文件。

```
<intent-filter>
    <action android:name="android.intent.action.MAIN" />
    <category android:name="android.intent.category.LAUNCHER" />
</intent-filter>
```

Data 描述 Intent 要操作的数据。以 URI 形式表示的数据如：content://contacts/1，可以通过 setData()方法实现 Data 属性的设置。

Type 用于指定该 Data 所指定的 Uri 对应 MIME 类型。这种 MIME 类型可以是任何自定义的 MIME 类型，只要符合 abc/xyz 格式的字符串即可。可以通过 setType()方法实现 Type 属性的设置。

Intent 的 Flag 属性用于为该 Intent 添加一些额外的控制旗标，Intent 可以调用 addFlags() 方法来为 Intent 添加控制旗标。

3. 常用按钮

Android 平台提供了很多视图组件，可以快速构建图形用户界面，常用的组件有文本组件、按钮组件、时间日期组件、列表组件、图像相关组件等。每种组件都需要相应的事件处理。事件处理的主要步骤是：实现监听器接口、覆盖监听方法、注册监听器。

ImageButton 是显示图像的按钮，是 ImageView 的子类。ImageView 组件用来在屏幕上显示图片，使用 android:src 属性设置 TextView 的图片源，或者调用 ImageView 类的 setImageResource(int resId)方法可以为 ImageView 设置显示的图片，本书任务 1.2 中详细介绍了 ImageView 中显示图片的方法。

```
<ImageView
    android:src="@drawable/moon"
    android:adjustViewBounds="true"
    android:layout_width="wrap_content"
    android:layout_height="wrap_content" />
<ImageButton
    android:layout_height="wrap_content"
    android:layout_width="wrap_content"
    android:src="@drawable/icon"/>
```

ImageSwitcher 可以用于以动画方式切换图像。ImageSwitcher 使用时需要一个工厂类创建在 ImageSwitcher 上显示的 View 对象，如 ImageView 对象。工厂类实现 ViewFactory 接口，覆盖 makeView 方法。

```
imageSwitcher = (ImageSwitcher) findViewById(R.id.imageswitcher);
imageSwitcher.setFactory(this);
imageSwitcher.setInAnimation(AnimationUtils.loadAnimation(this,
android.R.anim.fade_in));
imageSwitcher.setOutAnimation(AnimationUtils.loadAnimation(this,
android.R.anim.fade_out));
```

常用按钮 Button：如果按钮既需要显示图像又要显示文字，则使用<Button />设置。

```
<Button
    android:drawablePadding="20dp"
    android:drawableRight="@drawable/icon"
    android:layout_height="wrap_content"
    android:layout_width="wrap_content"
    android:text="拍照"/>
```

选项按钮 RadioButton：RadioButton 可以构建一组单选按钮，一组互斥的单选按钮必须在一个 RadioGroup 中。

```
</RadioGroup>
    android:layout_width="wrap_content">
<RadioButton
    android:layout_height="wrap_content"
    android:layout_width="wrap_content"  android:text="iPhone"/>
<RadioButton
    android:layout_height="wrap_content"
    android:layout_width="wrap_content"
    android:text="Android"/>
<RadioButton
    android:layout_height="wrap_content"
    android:layout_width="wrap_content"
    android:text="Phone"/>
</RadioGroup>
```

开关状态按钮 ToggleButton：ToggleButton 与 Button 的功能基本相同，ToggleButton 多了一个表示"开/关"状态的指示条。

```
<ToggleButton android:id="@+id/toggleButton"
    android:layout_height="wrap_content"
    android:layout_marginLeft="30dp"
    android:layout_width="wrap_content"
    android:textOff="已婚"
    android:textOn="未婚"/>
```

复选框按钮 CheckBox：CheckBox 多用于多选应用。

```
<CheckBox android:id="@+id/checkbox1"
    android:layout_height="wrap_content"
    android:layout_width="fill_parent"
    android:text="1、阅读"/>
<CheckBox android:id="@+id/checkbox2"
    android:layout_height="wrap_content"
    android:layout_width="fill_parent"
    android:text="2、旅游"/>
<CheckBox android:id="@+id/checkbox3"
    android:layout_height="wrap_content"
    android:layout_width="fill_parent"
    android:text="3、运动"/>
```

4．ListView

ListView 用于以垂直列表方式显示数据项。

ListView 使用步骤：

① 声明 ListView 布局文件，声明列表项的布局文件；

② 创建类继承 Activity 或 ListActivity；

③ 覆盖 onCreate()方法；

④ 创建 ListAdapter 对象，将 Adapter 对象设置到 ListView 中进行显示。

```
<ListView android:id="@+id/booklist"
    android:layout_height="wrap_content"
    android:layout_width="fill_parent"/>
```

ListView 的事件处理，与 ListView 列表项有关的事件有两种，可以使用 OnItemClickListener、OnItemSelectedListener 两个接口监听。如果 Activity 类继承了 ListActivity，则直接覆盖 onListItemClick 即可。

```
public void onItemClick(AdapterView<?> arg0, View arg1, int arg2, long arg3) {
    // TODO Auto-generated method stub
    Intent intent=new Intent(this,BookDetail.class);
    intent.putExtra("info", books[arg2].getDetail());
    this.startActivity(intent);
}
```

显示下拉列表 Spinner：Spinner 的功能和 ListView 组件类似，Spinner 的数据也是通过 Adapter 装载，使用数组或者 List 对象。

```
ArrayAdapter<String> adapter = new ArrayAdapter<String>(this,
    android.R.layout.simple_spinner_item, contactors);
spinner.setAdapter(adapter)
```

5．其他控件

输入日期/时间的组件 DatePicker/TimePicker。DatePicker 组件可以输入日期，范围是 1900-1-1 至 2100-12-31；TimePicker 组件可以输入时间，只能输入小时和分钟，默认情况是 12 小时制，对应的监听器分别是 OnDateChangedListener 和 OnTimeChangedListener。

显示时钟的组件 AnalogClock/DigitalClock：AnalogClock 用表盘方式显示当前时间，有时针和分针两个指针；DigitalClock 用数字方式显示当前时间，可以显示时、分、秒。

进度条 ProgressBar，用来显示任务或工作的完成率，Android 系统中可以实现圆形或者水平的进度条。ProgressBar 类中有 setProgress 和 getProgress 方法用来设置及获取当前进度，ProgressBar 类中有 setSecondaryProgress 和 getSecondaryProgress 方法用来设置及获取二级进度。

拖动条 SeekBar 是 ProgressBar 的子类，使用方式和 ProgressBar 类似，拖动条滑动的相关事件接口是 OnSeekBarChangerListener。该接口中 3 个方法：onProgressChanged 滑动滑杆、onStartTrackingTouch 按住滑杆和 onStopTrackingTouch 松开滑杆。

评分条 RatingBar，用来实现评分功能。RatingBar 常用的布局属性：android:numStars 用于评分的五角星个数，android:rating 指定当前的分数，android:stepSize 指定分数的增量单位，style 设置 RatingBar 的风格。

网格组件 GridView，以网格的形式排列所包含的内容，每个单元格的内容可以是任意一个 View 组件，与 ListView 相同。GridView 通过 ListAdapter 封装后台的数据，必须调用 setAdapter() 方法将 GridView 和数据绑定，GridView 可以使用 OnItemClickListener 及 OnItemSelectedListener 监听事件。

循环显示组件 Gallery，用于显示水平滚动的列表数据，其中心是固定不动的。Gallery 常常用来显示图像列表，也被称为相册组件。Gallery 和 GridView 的区别是 Gallery 只能显示一行，而且支持水平滑动效果。

标签组件 TabHost 是标签的集合，每一个标签是一个 TabHost.TabSpec 类的一个实例。TabHost 类的 addTab 方法可以添加多个 TabHost.TabSpec 实例，创建 TabHost 对象时，一般使用从 TabActivity 类继承。

任务实现

1．创建注册界面

在 res\layout 中创建 activity-regrster.xml，在 activity-regrster.xml 文件中根据页面要求设置 TextView、EditText、RadioGroup、RadioButton、ToggleButton、CheckBox、Spinner、Button 等控件。

```
<?xml version="1.0" encoding="utf-8"?>
<LinearLayout xmlns:android="http://schemas.android.com/apk/res/android"
    xmlns:tools="http://schemas.android.com/tools"
    android:layout_width="match_parent"
    android:layout_height="match_parent"
    android:background="@drawable/bg" >
```

```xml
<TableLayout
    android:layout_width="wrap_content"
    android:layout_height="wrap_content"
    android:collapseColumns="3"
    android:paddingLeft="25px"
    android:paddingRight="30px"
    android:paddingTop="30px" >
    <TableRow
        android:layout_width="wrap_content"
        android:layout_height="wrap_content" >
        <TextView
            android:id="@+id/textview01"
            android:layout_width="wrap_content"
            android:layout_height="wrap_content"
            android:text="用户名称" />
        <EditText
            android:id="@+id/username"
            android:layout_width="fill_parent"
            android:layout_height="wrap_content"
            android:text="" />
    </TableRow>
    <TableRow
        android:layout_width="wrap_content"
        android:layout_height="wrap_content" >
        <TextView
            android:layout_width="wrap_content"
            android:layout_height="wrap_content"
            android:text="用户密码" />
        <EditText
            android:id="@+id/password"
            android:layout_width="fill_parent"
            android:layout_height="wrap_content"
            android:inputType="textPassword"
            android:text="" />
    </TableRow>
    <TableRow
        android:layout_width="wrap_content"
        android:layout_height="wrap_content" >
```

```xml
<TextView
    android:layout_width="wrap_content"
    android:layout_height="wrap_content"
    android:text="性别" />
<RadioGroup
    android:id="@+id/gender_g"
    android:layout_width="wrap_content"
    android:layout_height="wrap_content" >
    <RadioButton
        android:id="@+id/male"
        android:layout_width="wrap_content"
        android:layout_height="wrap_content"
        android:text="男" />
    <RadioButton
        android:id="@+id/female"
        android:layout_width="wrap_content"
        android:layout_height="wrap_content"
        android:text="女" />
</RadioGroup>
</TableRow>
<TableRow
    android:layout_width="wrap_content"
    android:layout_height="wrap_content" >
    <TextView
        android:id="@+id/textviewmarriage"
        android:layout_width="wrap_content"
        android:layout_height="wrap_content"
        android:text="婚否" />
    <ToggleButton
        android:id="@+id/marriged"
        android:layout_width="wrap_content"
        android:layout_height="wrap_content"
        android:text="是" />
</TableRow>
<TableRow
    android:id="@+id/tablerow05"
    android:layout_width="wrap_content"
    android:layout_height="wrap_content" >
```

```xml
<TextView
    android:id="@+id/hobby"
    android:layout_width="wrap_content"
    android:layout_height="wrap_content"
    android:text="爱好" />
<LinearLayout
    android:layout_width="wrap_content"
    android:layout_height="wrap_content"
    android:gravity="left"
    android:orientation="horizontal" >
    <CheckBox
        android:id="@+id/reading"
        android:layout_width="wrap_content"
        android:layout_height="wrap_content"
        android:text="阅读" />
    <CheckBox
        android:id="@+id/swimming"
        android:layout_width="wrap_content"
        android:layout_height="wrap_content"
        android:text="游泳" />
    <CheckBox
        android:id="@+id/running"
        android:layout_width="wrap_content"
        android:layout_height="wrap_content"
        android:text="跑步" />
</LinearLayout>
</TableRow>
<TableRow
    android:id="@+id/tablerow06"
    android:layout_width="wrap_content"
    android:layout_height="wrap_content" >
    <TextView
        android:id="@+id/职务"
        android:layout_width="wrap_content"
        android:layout_height="wrap_content"
        android:text="职务" />
    <Spinner
        android:id="@+id/position"
```

```
                    android:layout_width="wrap_content"
                    android:layout_height="wrap_content" />
            </TableRow>
            <TableRow
        android:id="@+id/tablerow07"
        android:layout_width="wrap_content"
        android:layout_height="wrap_content" >
        <Button
            android:id="@+id/cancle"
            android:layout_width="wrap_content"
            android:layout_height="wrap_content"
            android:text="取消 " >
        </Button>
        <Button
            android:id="@+id/register"
            android:layout_width="wrap_content"
            android:layout_height="wrap_content"
            android:text="注册 " >
        </Button>
        </TableRow>
    </TableLayout>
</LinearLayout>
```

2. 实现注册功能

创建 RegisterActivity，在 RegisterActivity 中使用 findViewById()方法加载各项控件，并实现"注册"按钮的单击事件。在注册按钮事件中将注册信息以字符串形式写入 Bundle 对象当中，并通过 Intent 跳转到 ResultActivity 当中。

```
package com.siit.dict;
import android.app.Activity;
import android.os.Bundle;
import android.app.Activity;
import android.content.Intent;
import android.view.Menu;
import android.view.View;
import android.view.View.OnClickListener;
import android.widget.ArrayAdapter;
import android.widget.Button;
import android.widget.CheckBox;
import android.widget.EditText;
```

```java
import android.widget.RadioButton;
import android.widget.Spinner;
import android.widget.ToggleButton;
public class RegisterActivity extends Activity {
    //声明按钮
    private Button register,cancel;
    //声明ToggleButton
    private ToggleButton married;
    //声明单选按钮
    private RadioButton male,female;
    //声明文本编辑框
    private EditText username,password;
    //声明下拉列表
    private Spinner position;
    //声明多选按钮
    private CheckBox reading,swimming,running;
    @Override
    protected void onCreate(Bundle savedInstanceState) {
        super.onCreate(savedInstanceState);
        setContentView(R.layout.activity_register);
        username=(EditText)findViewById(R.id.username);
        password=(EditText)findViewById(R.id.password);
        male=(RadioButton)findViewById(R.id.male);
        female=(RadioButton)findViewById(R.id.female);
        reading=(CheckBox)findViewById(R.id.reading);
        swimming=(CheckBox)findViewById(R.id.swimming);
        running=(CheckBox)findViewById(R.id.running);
        married=(ToggleButton)findViewById(R.id.married);
        position=(Spinner)findViewById(R.id.position);
        String[] str={"CEO","CFO","PM"};
        ArrayAdapter aa=new ArrayAdapter(this, android.R.layout.simple_spinner_item,str);
        position.setAdapter(aa);
        register=(Button)findViewById(R.id.register);
        cancel=(Button)findViewById(R.id.cancle);
        register.setOnClickListener(new OnClickListener() {
            @Override
            public void onClick(View v) {
```

```java
        // TODO Auto-generated method stub
        Bundle b=new Bundle();
        b.putString("username", "用户名称"+username.getText().toString());
        b.putString("password", "用户密码"+password.getText().toString());
        if(male.isChecked()){
            b.putString("gender", "性别：男");
        }else{
            b.putString("gender", "性别：女");
        }
        String temp="爱好";
        if(reading.isChecked()){
            temp+="阅读";
        }
        if(swimming.isChecked()){
            temp+="游泳";
        }
        if(running.isChecked()){
            temp+="跑步";
        }
        b.putString("hobby",temp);
        if(marriged.isChecked()){
            b.putSerializable("marriged", "婚否：已婚");
        }
        else{
            b.putSerializable("marriged", "婚否：未婚");
        }
        b.putString("position", "职位"+position.getSelectedItem().toString());
        Intent intent=new Intent(RegisterActivity.this, ResultActivity.class);
        intent.putExtra("data", b);
        startActivity(intent);
        }
    });
    }
}
```

3. 实现登录页面向注册页面的跳转

在LoginAcitivity中实现注册按钮事件的响应，并利用Intent实现页面跳转。

在onCreate()方法中添加事件监听方法，代码如下。

```java
btnReg.setOnClickListener(this);
```

在 onClick(View arg0)方法中实现事件监听，代码如下。

```
    case R.id.regButton:
        Intent intent=new Intent();
        intent.setClass(LoginActivity.this, RegisterActivity.class);
        startActivity(intent);
        break;
```

4．创建结果页面

在 res\layout 中创建结果页面 result.xml，在该页面中设置 listView 控件。

```xml
<?xml version="1.0" encoding="utf-8"?>
<LinearLayout xmlns:android="http://schemas.android.com/apk/res/android"
    android:layout_width="match_parent"
    android:layout_height="match_parent"
    android:orientation="vertical" >
    <ListView
        android:id="@+id/listview"
        android:layout_width="wrap_content"
        android:layout_height="wrap_content" >
    </ListView>
</LinearLayout>
```

5．实现结果显示

创建 ResultActivity，用于显示用户注册结果，设置 activity_result 为显示界面，并加载 activity_result.xml 中的 listview 控件，通过 Intent 接收 LoginActivity 传递过来的信息，然后利用 ArrayAdapter 适配器显示用户注册信息。具体实现代码如下所示。

```java
import android.app.Activity;
import java.util.ArrayList;
import java.util.List;
import android.app.Activity;
import android.content.Intent;
import android.os.Bundle;
import android.widget.ArrayAdapter;
import android.widget.ListView;
public class ResultActivity extends Activity {
    private ListView listview;
    protected void onCreate(Bundle savedInstanceState) {
        super.onCreate(savedInstanceState);
        setContentView(R.layout.activity_result);
        listview=(ListView)findViewById(R.id.listview);
```

```
            Intent intent=this.getIntent();
            Bundle b=intent.getBundleExtra("data");
            List list=new ArrayList();
            list.add(b.getString("username"));
            list.add(b.getString("password"));
            list.add(b.getString("position"));
            list.add(b.getString("gender"));
            list.add(b.getString("hobby"));
            list.add(b.getString("married"));
            ArrayAdapter adapter=new ArrayAdapter(this, android.R.layout.simple_spinner_item,list);
            listview.setAdapter(adapter);
    }
}
```

6. 配置 AndroidManifest.xml 文件

在 AndroidMainfestxml 注册 RegisterActivity 与 ResultActivity。在 AndroidMainfestxml 文件的<application>元素中添加两个<activity>元素，内容如下。

```
<activity android:name="com.siit.dict.RegisterActivity"/>
<activity android:name="com.siit.dict.ResultActivity"/>
```

任务 2.4　用户访问信息存取

任务描述

在 dict 工程中编程实现用户登录访问信息的存取，通过 Sharedpreferences 方式实现用户名、登录次数和最近登录时间的存取，效果如图 2-17 所示。

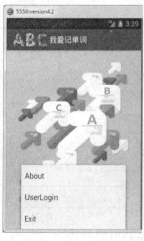

图 2-17　程序运行效果

任务目标

① 理解 SharedPreferences 的概念。
② 掌握使用 SharedPreferences 的常用方法掌握 File 存取的方法。
③ 会利用 SharedPreferences 与 File 实现数据的写入和存取。

任务分析

本任务主要实现将用户登录信息的写入 SharedPreferences 文件中,并通过菜单单击事件从 SharedPreferences 文件中读取用户信息,并显示用户登录信息,具体实现过程如下。

① 将用户登录信息写入 SharedPreferences。
② 从 SharedPreferences 文件中读取用户登录信息。
③ 实现从 LoginActivity 到 UserloginInfoActivity 的跳转。
④ 配置 AndroidManifest.xml 文件。

知识要点

Android 平台提供的 SharedPreferences 类是一个轻量级的存储类,适用于保存软件配置参数。此外 Android 也可以像 Java 程序一样存入和读取 File 文件的内容,下面介绍这两类文件的操作方法。

1. SharedPreferences 文件管理

使用 SharedPreferences 保存数据,其背后使用 xml 文件存放数据,文件存放在 /data/data/<package name>/shared_prefs 目录下。

(1) sharedPreferences 文件的创建与存取

有 3 种方法可以获取 Preferences 文件,下面来分别介绍。

- this.getPreferences (int mode):通过 Activity 对象获取,获取的是本 Activity 私有的 Preference,保存在系统中的 xml 形式的文件的名称为这个 Activity 的名字,因此一个 Activity 只能有一个 Preferences 文件属于该 Activity。Activity 提供了 getPreferences(mode) 方法操作 SharedPreferences,这个方法默认使用当前类不带包名的类名作为文件的名称。
- PreferenceManager.getDefaultSharedPreferences(this):PreferenceManager 的静态函数,保存 PreferenceActivity 中的设置,属于整个应用程序,但是只有一个 Android 会根据包名和 PreferenceActivity 的布局文件来起一个名字保存。
- this.getSharedPreferences (String name, int mode):因为 Activity 继承了 ContextWrapper,因此也是通过 Activity 对象获取,但是属于整个应用程序,可以有多个。SharedPreferences 通过 Activity 自带的 getSharedPreferences(name,mode)方法指定文件,如果是第一次运行则先创建该文件。getSharedPreferences(name,mode)方法的第一个参数用于指定该文件的名称,名称不用带后缀,后缀会由 Android 自动加上;方法的第二个参数指定文件的操作模式,共有 4 种操作模式。

SharePreferences 操作文件的 4 种模式：
- Context.MODE_PRIVATE=0
- Context.MODE_APPEND=32768
- Context.MODE_WORLD_READABLE=1
- Context.MODE_WORLD_WRITEABLE=2

Context.MODE_PRIVATE：默认操作模式，代表该文件是私有数据，只能被应用本身访问，在该模式下写入的内容会覆盖原文件的内容。

Context.MODE_APPEND：模式会检查文件是否存在，存在就往文件追加内容，否则就创建新文件。如果想把新写入的内容追加到原文件中，可以使用 Context.MODE_APPEND。

Context.MODE_WORLD_READABLE 和 Context.MODE_WORLD_WRITEABLE 用来控制其他应用是否有权限读写该文件。

MODE_WORLD_READABLE：表示当前文件可以被其他应用读取。

MODE_WORLD_WRITEABLE：表示当前文件可以被其他应用写入。

如果希望 SharedPreferences 背后使用的 xml 文件能被其他应用读和写，可以指定 Context.MODE_WORLD_READABLE 和 Context.MODE_WORLD_WRITEABLE 权限。

（2）SharedPreferences 文件内容的编辑与存取

SharedPreferences 类中的 edit()方法用于对 SharedPreferences 文件内容进行编辑。Editor 类 putXXX()方法用于数据的存入，使用 Preferences 的键值对存储方式，这种方式主要用来存储比较简单的一些数据，而且是标准的 Boolean、Int、Float、Long、String 等类型。

Editor 类的 commit()方法用于向 sharedPreferences 提交修改。

下面的代码实现将一个随机数的存入到 sharedPreferences 文件当中。

```
SharedPreferences sharedPreferences = getSharedPreferences("dict", Context.MODE_PRIVATE);
//该语句在创建和读取文件时可以只声明一次，用于打开或创建 dict.xml 文件
Editor editor = sharedPreferences.edit();//获取编辑器
editor.putInt("random",(int)(Math.random()*100));
editor.commit();//提交修改
```

生成的 dict.xml 文件内容如下。

```
<?xml version='1.0' encoding='utf-8' standalone='yes' ?>
<map>
<int name="random" value="73" />
</map>
```

SharedPreferences 文件内容的读取，首先指定并打开 Preferences 文件,然后通过 getXXX()方法读取文件内容，代码示例如下。

```
SharedPreferences preferences=getSharedPreferences("dict",MODE_PRIVATE);
//该语句在创建和读取文件时可以只声明一次，用于打开或创建 dict.xml 文件
int randnum=preferences.getInt("random", 0);
```

2. Android 中使用 File 文件

Android 中使用 File 文件进行数据存储。有时候可以将数据直接以文件的形式保存在设备中，如文本文件、图片文件等。使用 File 进行存储操作主要使用以下方法。

- public abstract FileInputStream openFileInput (String name)：主要是打开文件，返回 FileInputStream。
- public abstract FileOutputStream openFileOutput (String name, int mode)：主要是写入文件，如果该文件不存在，直接进行创建，返回 FileOutputStream。

Mode 主要有 4 种模式：MODE_APPEND 在尾部追加、MODE_PRIVATE 私有、MODE_WORLD_READABLE 可读、MODE_WORLD_WRITEABLE 可写。

FileInputStream（获取文件输入流）与 FileOutputStream （获取文件输出流）这两类在 JavaIO 操作中很常见，接下来进行操作保存成功之后，文件保存在当前应该程序的包名下的 files/（可以改变存储的其他路径）如图 2-18 所示，具体代码参见任务实现内容。

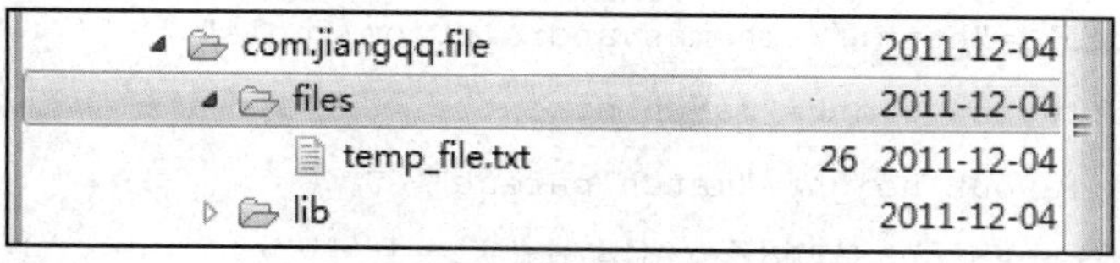

图 2-18 File 文件存储路径图

任务实现

1．将用户登录信息写入 SharedPreferences

用户单击"登录"按钮时，将用户登录信息写入 SharedPreferences。用户登录信息包括用户名、登录次数、最近登录时间。

① 编写 saveLoinginfor()方法实现将用户登录信息写入 SharedPreferences。

```
void saveLoinginfor(){
    SharedPreferences preferences;
    SharedPreferences.Editor editor;
    preferences=getSharedPreferences("login",MODE_APPEND);
    editor=preferences.edit();
    int count=0;
    //写入用户登录的时间与次数
    SimpleDateFormat sdf=new SimpleDateFormat("yyyy年MM月dd日"+"hh:mm:ss");
    editor.putString("time",sdf.format(new Date()));
    editor.putString("username", username.getText().toString());
    count=preferences.getInt("count", 0);
    editor.putInt("count", ++count);
    editor.commit();
```

② 在"登录"按钮的单击事件中调用 saveLoinginfor()实现用户信息写入。相关代码如下。

```
public void onClick(View v) {
case R.id.loginButton:
……
    saveLoinginfor();
……
}
```

2. 从 SharedPreferences 中读取用户登录信息

① 在 res\layout 中创建 activity_userloginfo.xml，实现用户登录信息显示页面设计。

```xml
<?xml version="1.0" encoding="utf-8"?>
<LinearLayout xmlns:android="http://schemas.android.com/apk/res/android"
    xmlns:tools="http://schemas.android.com/tools"
    android:layout_width="match_parent"
    android:layout_height="match_parent"
    android:background="@drawable/bg"
    android:orientation="vertical"
    tools:context=".UserloginActivity" >
    <LinearLayout
        android:layout_width="match_parent"
        android:layout_height="match_parent"
        android:orientation="vertical"
        android:paddingLeft="35px"
        android:paddingRight="40px"
        android:paddingTop="40px" >
        <TextView
            android:id="@+id/username"
            android:layout_width="wrap_content"
            android:layout_height="wrap_content" />
        <TextView
            android:id="@+id/loingcount"
            android:layout_width="wrap_content"
            android:layout_height="wrap_content" />
        <TextView
            android:id="@+id/logintime"
            android:layout_width="wrap_content"
            android:layout_height="wrap_content" />
```

```
        </LinearLayout>
</LinearLayout>
```

② 创建 UserloginInfoActivity,将 activity_userloginfo.xml 文件中的控件使用 findViewById() 方法加载到该 Activity 当中,并读取 SharedPreferences 文件中信息,如用户名、登录次数、最近登录时间,将其显示到页面当中。

```java
import java.text.SimpleDateFormat;
import java.util.ArrayList;
import java.util.Date;
import java.util.List;
import android.os.Bundle;
import android.app.Activity;
import android.content.SharedPreferences;
import android.view.Menu;
import android.widget.ArrayAdapter;
import android.widget.ListView;
import android.widget.TextView;
public class UserloginInfoActivity extends Activity {
    int count=0;
    private TextView username,loginCount;
    private TextView loginTime;
    @Override
    protected void onCreate(Bundle savedInstanceState) {
        super.onCreate(savedInstanceState);
        setContentView(R.layout.activity_userloginfo);
        username=(TextView) findViewById(R.id.username);
        loginCount=(TextView) findViewById(R.id.loingcount);
        loginTime=(TextView) findViewById(R.id.logintime);
        readLoinginfor();
    }
    void readLoinginfor(){
        SharedPreferences preferences;
        SharedPreferences.Editor editor;
        preferences=getSharedPreferences("login",MODE_PRIVATE);
        editor=preferences.edit();
        String usernametext=preferences.getString("username", null);
        username.setText("用户名: \t"+usernametext);
        count=preferences.getInt("count",0);
        String time=preferences.getString("time", null);
```

```
        loginCount.setText("登录次数：\t"+String.valueOf(count));
        loginTime.setText("最近登录时间：\t"+String.valueOf(time));
    }
}
```

3．实现用户登录页面到用户访问信息页面的跳转

在 LoginActivity 中实现菜单事件，先修改 aboutuserinfo(String msg)方法，在方法中实现 Intent 跳转，并在用户登录页面的菜单事件中调用该方法，相关代码如下。

```
    private void aboutuserinfo(String msg){
        AlertDialog.Builder builder = new AlertDialog.Builder(this);
        builder.setMessage(msg)
        .setCancelable(false)
        .setPositiveButton("确定", new DialogInterface.OnClickListener() {
            public void onClick(DialogInterface dialog, int id) {
                Intent intent=new Intent();
                intent.setClass(LoginActivity.this, UserloginInfoActivity.class);
                startActivity(intent);
            }
        });
        AlertDialog alert = builder.create();
        alert.show();
    }
    @Override
    public boolean onOptionsItemSelected(MenuItem item) {
        switch (item.getItemId()) {
        case R.id.about:
            aboutAlert("这是一个用于单词记忆和管理的小程序！");
            break;
        case R.id.exit:
            exitAlert("真的要退出吗？");
            break;
        case R.id.userLogin:
            aboutuserinfo("用户访问信息查看！");
            break;
        }
        return true;
    }
```

4．配置 AndroidManifest.xml 文件

`<activity android:name="com.siit.dict.UserloginInfoActivity"/>`

技能训练

通过 SharePreferences 实现当前时间、一个随机数和登录次数的存取。通过 File 文件实现数据存取如图 2-19 所示。

在 Activity 中生成当前时间、一个随机数和登录次数，单击"写入"按钮通过 Shareferences 进行数据保存，单击"读取"按钮是从 Shareferences 中获取写入时间、生成的随机数和登录次数，并使用 Toast 将信息显示。

从 EditText 中写入数据，单击"写入"按钮时将 EditText 文本框中的文字保存到 File 当中去，单击"读取"按钮将 File 文件的内容读取，并显示到 TextView 控件中。

图 2-19　程序运行效果

任务 2.5　单词存取

任务描述

利用 SQLiteDatabase 和 SQLiteOpenHelper 类实现单词信息的添加、删除、修改和查询。其中一个界面实现单词管理，包括单词的添加、更新、删除，"单击查看"后，另一个界面用列表显示单词数据，其效果如图 2-20 所示。

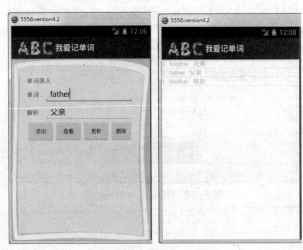

图 2-20　程序运行效果

任务目标

① 掌握 SQLiteDatabase 数据库和数据表的创建。

② 掌握 SQLiteDatabase 类的常用方法。
③ 理解 SQLiteOpenHelper 类的概念。
④ 掌握 SQLiteOpenHelper 类的常用方法。

任务分析

本任务主要实现单词信息的管理，包括单词的添加、删除、修改和查询。单词的数据表结构如图 2-21 所示，任务完成后的数据如图 2-22 所示。具体实现过程如下所示：

① 准备字符串资源文件；
② 数据层设计与实现；
③ 实现单词管理界面；
④ 实现单词管理；
⑤ 实现单词查询结果显示；
⑥ 实现登录页面向单词管理页面的跳转；
⑦ 配置 AndroidManifest.xml 文件。

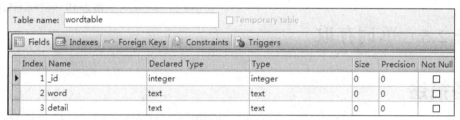

图 2-21　SQLite 中 wordtable 表结果

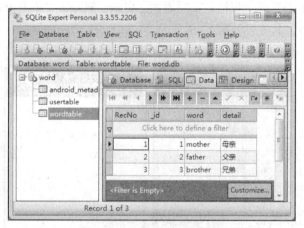

图 2-22　SQLite 中数据结果

知识要点

SQLite 是一款轻量级的关系型数据库。由于它占用的资源非常少，所以 Android 系统中用 SQLite 来存储数据。SQLiteDatabase 与 SQLiteOpenHelper 是 Android 系统中用于管理 SQLite 数据库的类。

1. SQLiteDatabase 简介

Android 平台内置了对 SQLite 数据库，它是轻量级的嵌入式数据库。对于数据结构较为复杂的关系型数据，使用 SQLite 存储将非常高效。应用程序创建的数据库文件存储在 /data/data/包名/database 目录下，应用程序之间不能相互访问。

（1）创建 SQLiteDatabase 数据库

Context.openOrCreateDatabase(String name,int mode,CursorFactory factory)

- name 代表数据库的名称。
- mode 代表创建数据库的模式，包括 MODE_PRIVATEMODE_WORLD_READABLE 和 MODE_WORLD_WRITEABLE。

使用 openOrCreateDatabase()之前要首先查询数据库是否已经存在，效率较低。

（2）创建 SQLite 数据表的方法

execSQL(create table _id INTEGER primary key,word text,detail text);

（3）SQLiteDatabase 数据插入

方法一：insert(databaseHelper.DATABASE_TABLE, null, content)。

参数说明：

- 第一个参数 databaseHelper.DATABASE_TABLE 数据库表名；
- 第二个参数如果 content 为空时则向表中插入一个 null；
- 第三个参数为插入的内容。

方法二：execSQL(String sql)，sql 为实现数据插入的 SQL 命令字符串。

（4）SQLiteDatabase 数据删除

方法一：delete(databaseHelper.DATABASE_TABLE,KEY_ROWID +"="+rowId , null)。

- 第一参数 databaseHelper.DATABASE_TABLE 为数据库表名。
- 第二个参数 KEY_ROWID +"="+rowId 表示条件语句。
- 第三个参数可表示为条件带?号的表达式。

delete()方法返回值大于 0 表示删除成功。

方法二：execSQL(String sql)，sql 为实现数据删除的 SQL 命令字符串。

（5）SQLiteDatabase 数据修改

方法一：update(databaseHelper.DATABASE_TABLE, args, KEY_ROWID + "=" + rowId, null)。

- 第一个参数 databaseHelper.DATABASE_TABLE 表示数据库表名。
- 第二个参数 args 表示更新的内容。
- 第三个参数 KEY_ROWID + "=" + rowId 表示更新的条件。
- 第四个参数可表示为条件带?号的表达式。

update()方法返回值大于 0 表示删除成功。

方法二：execSQL(String sql)，sql 为实现数据修改的 SQL 命令字符串。

（6）SQLiteDatabase 数据查询

SQLiteDatabase 提供了 Cursor query(String table, String[] columns, String selection, String[] selectionArgs, String groupBy, String having, String orderBy)方法来实现查询记录。

参数介绍：

- table 代表查询的数据表；
- columns 代表要查询的列数组；
- selection 代表查询的条件，可以使用"？"通配符；
- selectionArgs 代表 selection 表达式中的?号；
- groupBy 代表数据分组；
- having 代表哪些行显示；
- orderBy 代表排序的方式。

2．Cursor 类

Cursor 类用于对数据库查询的结果进行随机的读写访问。默认情况下 Cursor 的游标位于返回的所有数据行的前面。

Cursor 常用方法如下。

- move(int offset)：将当前游标移动 offset 个位置。
- moveToFirst()：将游标移动到第一行。
- moveToLast()：将游标移动到最后一行。
- isLast()：判断游标是否在最后一行。
- isFirst()：判断游标是否在第一行。
- getPosition()：获得游标当前所在行的位置。
- getCount()：获得 cursor 中的所有行数。

3．SQLiteOpenHelper 类

使用 SQLiteOpenHelper 创建数据库不会重复执行数据库的初始化操作，不需要查询数据库是否存在，执行效率更高。SQLiteOpenHelper 是一个抽象类，需要实现如下方法。

- onCreate(SQLiteDatabase db)
- onUpgrade(SQLiteDatabase db,int oldVersion,int newVersion)

调用 SQLiteOpenHelper 类的 getReadableDatabase()方法可以获得一个只读的 SQLiteDatabase 对象。

SQLiteDatabase db = SQLOpenHelper.getReadableDatabase();

调用 SQLiteOpenHelper 类的 getWritableDatabase()方法可以获得可写的 SQLiteDatabase 对象。

SQLiteDatabase db = SQLOpenHelper.getWriteableDatabase();

4．SimpleCursorAdapter 简介

public SimpleCursorAdapter (Context context, int layout, Cursor c, String[] from, int[] to,int flags)为该适配器类标准的构造函数。

参数如下所述。

- context：应用程序上下文，具体来说就是 ListView 所在的上下文档中。
- layout：布局文件的资源定位标识符，也就是说标识了 ListView 中的 item，那么这个布局文件至少包含了参数"to"中的传进来值。
- c：数据库游标，如果游标不可用则为 null。
- from：列名字列表，表示要绑定到 UI 上的列。如果游标不可用则为 null。

- to：展示参数"from"中的列，也就是说 ListView 中的视图显示的是参数"from"的列值，这些视图应该都是 TextView。如果游标不可用则为 null。
- flags：这个标志用来决定该适配器的行为。设置标志用来添加一个监听器，监听着参数 cursor 的数据是否有改变。

任务实现

1. 准备字符串资源文件

在 res/values/strings.xml 中添加以下字符串资源，供程序调用。

```xml
<string name="app_name">我爱记单词</string>
<string name="title">单词录入</string>
<string name="wordshow">单词查看</string>
<string name="menu_settings">Settings</string>
<string name="btninsert">添加</string>
<string name="btnSave">保存</string>
<string name="btnsearch">查看</string>
<string name="btnupdate">更新</string>
<string name="btndelete">删除</string>
<string name="word">单词：</string>
<string name="detail">解析：</string>
<string name="btnprior">前一条</string>
<string name="btnnext">后一条</string>
<string name="editWord">编辑单词</string>
<string name="updateword">单词重新 下载</string>
```

2. 数据层设计与实现

创建 WordDBHelper，该类继承于 SQLiteOpenHelper 类。通过 SQLiteDateBase 与 SQLiteOpenHelper 实现数据库的创建、表的创建，定义数据的增加、删除、修改和查询等方法。

```java
import android.content.ContentValues;
import android.content.Context;
import android.database.Cursor;
import android.database.sqlite.SQLiteDatabase;
import android.database.sqlite.SQLiteDatabase.CursorFactory;
import android.database.sqlite.SQLiteOpenHelper;
public class WordDBHelper extends SQLiteOpenHelper {
    private SQLiteDatabase db;
    private static final String CREATE_SQL="create table wordtable" +
            "(_id integer primary key autoincrement,word text,detail
```

```java
text)";
    public WordDBHelper(Context context) {
        super(context, "word.db", null, 2);
    }
    @Override
    public void onCreate(SQLiteDatabase db) {
        this.db=db;
        db.execSQL(CREATE_SQL);
    }
    @Override
    public void onUpgrade(SQLiteDatabase arg0, int arg1, int arg2) {
        // TODO Auto-generated method stub
    }
    public int insertDB(String word,String detail){
        db=getWritableDatabase();
        int i=0;
        ContentValues cv=new ContentValues();
        cv.put("detail",detail);
        cv.put("word",word);
        i=(int) db.insert("wordtable", null, cv);
        db.close();
        return i;
    }
    public int deleteDB(String word){
        db=getWritableDatabase();
        int i=0;
        String[] args={word};
        if(db==null)
            db=getReadableDatabase();
        i=db.delete("wordtable", "word=?",args);
        return i;
    }
    public Cursor query(){
        db=getWritableDatabase();
        Cursor c=db.query("wordtable", null, null, null, null, null, null);
        return c;
    }
    public int updateDB(String word,String detail){
```

```
        db=getWritableDatabase();
        int i=0;
        ContentValues cv=new ContentValues();
        cv.put("word", word);
        cv.put("detail", detail);
        String[] args=new String[1];
        args[0]=word;
        i=db.update("wordtable", cv, "word=?", args);
        return i;
    }
    public void close(){
        if(db!=null)
            db.close();
    }
}
```

3. 单词管理界面

使用 activity_word.xml 实现单词管理界面的设计。

```
<?xml version="1.0" encoding="utf-8"?>
<LinearLayout xmlns:android="http://schemas.android.com/apk/res/android"
    xmlns:tools="http://schemas.android.com/tools"
    android:layout_width="match_parent"
    android:layout_height="match_parent"
    android:gravity="left"
    android:orientation="vertical"
    tools:context=".MainActivity" >
    <TextView
        android:layout_width="wrap_content"
        android:layout_height="wrap_content"
        android:layout_centerHorizontal="true"
        android:layout_centerVertical="true"
        android:text="@string/title"
        android:textStyle="normal" />
    <TableRow
        android:layout_width="wrap_content"
        android:layout_height="wrap_content" >
        <TextView
            android:layout_width="wrap_content"
            android:layout_height="wrap_content"
```

```xml
            android:layout_centerHorizontal="true"
            android:layout_centerVertical="true"
            android:text="@string/word" />
    <EditText
        android:id="@+id/word"
        android:layout_width="fill_parent"
        android:layout_height="wrap_content"
        android:width="150dp" />
</TableRow>
<TableRow
    android:layout_width="wrap_content"
    android:layout_height="wrap_content" >
    <TextView
        android:layout_width="wrap_content"
        android:layout_height="wrap_content"
        android:layout_centerHorizontal="true"
        android:layout_centerVertical="true"
        android:text="@string/detail" />
    <EditText
        android:id="@+id/detail"
        android:layout_width="fill_parent"
        android:layout_height="wrap_content"
        android:width="150dp" />
</TableRow>
<TableRow
    android:layout_width="wrap_content"
    android:layout_height="wrap_content" >
    <Button
        android:id="@+id/btninsert"
        android:layout_width="wrap_content"
        android:layout_height="wrap_content"
        android:text="@string/btnok" />
    <Button
        android:id="@+id/btnsearch"
        android:layout_width="wrap_content"
        android:layout_height="wrap_content"
        android:text="@string/btnsearch" />
    <Button
```

```xml
            android:id="@+id/btnupdate"
            android:layout_width="wrap_content"
            android:layout_height="wrap_content"
            android:text="@string/btnupdate" />
        <Button
            android:id="@+id/btndelete"
            android:layout_width="wrap_content"
            android:layout_height="wrap_content"
            android:text="@string/btndelete" />
    </TableRow>
</LinearLayout>
```

4. 实现单词管理

使用 WordManageActivity 实现界面显示，获取文本值并对相应的按钮进行事件响应。

```java
import android.os.Bundle;
import android.annotation.SuppressLint;
import android.app.Activity;
import android.app.ListActivity;
import android.content.ContentValues;
import android.content.Intent;
import android.database.Cursor;
import android.database.sqlite.SQLiteDatabase;
import android.view.Menu;
import android.view.View;
import android.view.View.OnClickListener;
import android.widget.Button;
import android.widget.EditText;
import android.widget.ListView;
import android.widget.SimpleCursorAdapter;
import android.widget.Toast;
public class WordManageActivity extends Activity implements OnClickListener {
    private Button btninsert,btnupdate,btndelete,btnsearch;
    private EditText editTextword,editTextdetail;
    private String word,detail;
    private int i;
    WordDBHelper dbhelper;
    @Override
    protected void onCreate(Bundle savedInstanceState) {
        super.onCreate(savedInstanceState);
```

```java
        setContentView(R.layout.activity_main);
        btninsert=(Button)findViewById(R.id.btninsert);
        btnupdate=(Button)findViewById(R.id.btnupdate);
        btndelete=(Button)findViewById(R.id.btndelete);
        btnsearch=(Button)findViewById(R.id.btnsearch);
        editTextword=(EditText)findViewById(R.id.word);
        editTextdetail=(EditText)findViewById(R.id.detail);
        dbhelper=new WordDBHelper(this);
        btninsert.setOnClickListener(this);
        btnupdate.setOnClickListener(this);
        btndelete.setOnClickListener(this);
        btnsearch.setOnClickListener(this);
    }
    @Override
    public boolean onCreateOptionsMenu(Menu menu) {
    // Inflate the menu; this adds items to the action bar if it is present.
        getMenuInflater().inflate(R.menu.activity_main, menu);
        return true;
    }
    @SuppressLint("ShowToast")
    public void onClick(View arg0) {
        // TODO Auto-generated method stub
        switch(arg0.getId()){
        case R.id.btninsert:
            word=editTextword.getText().toString().trim();
            detail=editTextdetail.getText().toString().trim();
            i=dbhelper.insertDB(word,detail);
            if(i!=0)
                Toast.makeText(MainActivity.this, "单词录入成功 ", 8000).show();
            else
                Toast.makeText(MainActivity.this, "单词录入失败 ", 8000).show();
            break;
        case R.id.btnupdate:
            word=editTextword.getText().toString().trim();
            detail=editTextdetail.getText().toString().trim();
            i=dbhelper.updateDB(word,detail);
```

```
            if(i!=0)
                Toast.makeText(MainActivity.this,"单词修改成功 ",8000).show();
            else
                Toast.makeText(MainActivity.this,"单词修改失败 ",8000).show();
            break;
        case R.id.btnsearch:
            Intent intent=new Intent(MainActivity.this,QueryActivity.class);
            startActivity(intent);
            break;
        case R.id.btndelete:
            word=editTextword.getText().toString().trim();
            i=dbhelper.deleteDB(word);
            if(i!=0)
                Toast.makeText(MainActivity.this,"单词删除成功 ",8000).show();
            else
                Toast.makeText(MainActivity.this,"单词删除失败 ",8000).show();
            break;
        }
    }
}
```

5．实现单词查询结果显示

①界面设计 row.xml 用于实现查询内容的显示。

```xml
<?xml version="1.0" encoding="utf-8"?>
<LinearLayout xmlns:android="http://schemas.android.com/apk/res/android"
    android:orientation="horizontal" android:layout_width="fill_parent"
    android:layout_height="fill_parent" android:layout_gravity="center_vertical">
    <TextView
    android:id="@+id/textid"
    android:layout_width="wrap_content"
    android:layout_height="wrap_content"
    android:paddingRight="10px"/>
    <TextView
    android:id="@+id/textword"
    android:layout_width="wrap_content"
```

```xml
        android:layout_height="wrap_content"
        android:paddingRight="10px" />
    <TextView
        android:id="@+id/textdetail"
        android:layout_width="wrap_content"
        android:layout_height="wrap_content"
        android:paddingRight="10px"/>
</LinearLayout>
```

② 使用 QueryActivity.java 实现查询数据的显示。

```java
import android.app.AlertDialog;
import android.app.ListActivity;
import android.content.DialogInterface;
import android.database.Cursor;
import android.os.Bundle;
import android.view.View;
import android.widget.AdapterView;
import android.widget.ListView;
import android.widget.SimpleCursorAdapter;
import android.widget.AdapterView.OnItemClickListener;
public class QueryActivity extends ListActivity {
    @Override
    protected void onCreate(Bundle savedInstanceState) {
        // TODO Auto-generated method stub
        final WordDBHelper dbhelper;
        super.onCreate(savedInstanceState);
        dbhelper=new WordDBHelper(this);
        Cursor c = dbhelper.query();
        String[] from = {"_id", "word", "detail"};
        int[] to = { R.id.textid,R.id.textword, R.id.textdetail};
        SimpleCursorAdapter adapter = new SimpleCursorAdapter(getApplicationContext(),
R.layout.row, c, from, to);
        ListView listView = getListView();
        listView.setAdapter(adapter);
        dbhelper.close();
    }
}
```

6. 实现登录页面向单词管理页面的跳转

在 LoginAcitivity 登录按钮的单击事件中利用 Intent 实现向单词管理页面的跳转，在 onClick(View arg0)方法中的 case R.id.regButton:中实现事件监听，添加如下代码。

```
…
        Intent intent1=new Intent();
        intent1.setClass(LoginActivity.this, WordManageActivity.class);
        startActivity(intent1);
…
```

7. 配置 AndroidManifest.xml 文件

```
<activity android:name="com.siit.dict.WordManageActivity"/>
<activity android:name="com.siit.dict.QueryActivity"/>
```

技能训练

1. 结合 SQLite 实现用户查询与用户注册功能

界面设计如图 2-23 所示。

图 2-23　程序运行效果

数据表创建：在 WordDBHelper 中通过 CREATE_SQLuser 定义"用户信息表"的创建语句，并在 onCreate 方法中使用 db.execSQL(CREATE_SQLuser)实现用户信息表的创建，用户数据表的结构如图 2-24 所示，其数据添加成功后数据表中内容如图 2-25 所示。

注册功能：在 WordDBHelper 中定义方法 int insertUser(String username, String password, String ssex,boolean marrage, String favorite,String position)，定义 Content cv 对象，通过 SQLite 的 db.insert("usertable", null, cv)方法实现用户信息的注册，添加成功显示用户注册成功。

登录功能：在 WordDBHelper 中定义方法 Cursor queryUserByname(String username,String password)通过 SQLite 的 query()方法验证用户名和密码是否正确，存在该用户名和密码则显示

登录成功，否则显示"用户名或密码不正确"。

在登录和注册页面中分别调用 insertUser()和 queryUserByname()方法实现用户登录验证和用户注册。

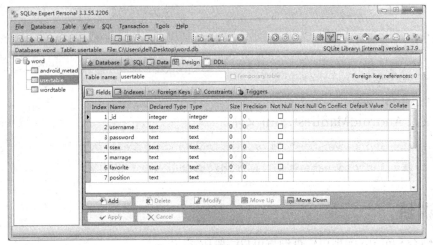

图 2-24 用户数据表结构

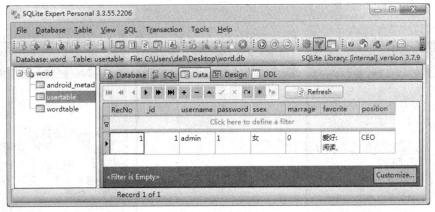

图 2-25 用户信息

2．利用 SQLite 数据库的查询功能实现单词查看

默认显示单词信息的第一条数据，单击"上一条"和"下一条"实现单词信息的滚动显示。界面效果如图 2-26 所示，步骤分析如下。

- 在 WordDBHelper 中定义 query()方法，用于返回单词信息。
- 在 WordShowActivity 通过 WordDBHelper 对象调用 query()方法实现单词信息的查询，通过 moveToFirs()显示第一条数据，moveToPrevious()查看上一条，moveToNext()查看下一条数据。
- 单击"编辑单词"跳转到 WordManageActivity 页面。

图 2-26 单词查看运行效果

任务 2.6 单词共享

任务描述

使用 ContentProvider 将 dict 工程中的单词信息定义为内容提供者,共享给其他应用程序,使得其他应用程序可以访问 dict 工程当中的数据。

创建 CPTest 工程,在 CPTest 应用程序中通过 ContentProvider 读取 dict 工程中的单词信息,并逐条显示。

功能效果如图 2-27 和图 2-28 所示。

图 2-27 dict 单词管理   图 2-28 CPTest 访问 dict 的单词效果

任务目标

① 了解 ContentProvider 的概念。

② 知道 ContentProvider 的作用。
③ 掌握 ContentProvider 的声明与定义过程。
④ 能熟练使用 ContentProvider 实现单词的添加、删除、修改。
⑤ 能实现不同应用间的内容共享。

知识要点

1. ContentProvider 简介

ContentProvider 在 Android 系统中的作用是对外共享数据，可以通过 ContentProvider 把应用中的数据共享给其他应用访问，其他应用可以通过 ContentProvider 对应用中的数据进行增、删、改、查。

定义 ContentProvider 时需要定义一个类继承自 ContentProvider 类，并重写该类的方法。虽然使用其他方法也可以对外共享数据，但数据访问方式会因数据存储的方式而不同。例如：采用文件方式对外共享数据，需要进行文件操作读写数据；采用 sharedpreferences 共享数据，需要使用 sharedpreferences API 读写数据。而使用 ContentProvider 共享数据的好处是统一了数据访问方式。

对于 ContentProvider 基本操作分为两个部分，第一定义 ContentProvider，第二使用 ContentProvider。其中 Uri 是 ContentProvider 使用过程中的重要概念。

2. Uri 的定义与解析

在 ContentProvider 的定义和使用之间利用 Uri 进行传递操作指令，所以先介绍 Uri 的概念，便于理解 ContentProvider 的定义和使用。Uri 代表了要操作的数据，主要包含了两部分信息。

① 需要操作的 ContentProvider。
② 对 ContentProvider 中的什么数据进行操作。

一个 Uri 由以下几部分组成。

- ContentProvider（内容提供者）的 Scheme 已经由 Android 所规定，Scheme 为 content://主机名（或叫 Authority），用于唯一标识这个 ContentProvider，外部调用者可以根据这个标识来找到它。
- 路径（path）可以用来表示要操作的数据，路径的构建应根据业务而定。

要操作 word 表中 id 为 10 的记录，可以构建这样的路径:/word/10。
要操作 word 表中 id 为 10 的记录的 name 字段，可以构建路径：word/10/name 。
要操作 word 表中的所有记录，可以构建路径:/word。
要操作 xxx 表中的记录，可以构建路径:/xxx。
要操作的数据不一定只有数据库文件，也可以是文件、xml 或网络等其他形式的内容。
要操作 xml 文件中 word 节点下的 name 节点，可以构建路径：/word/name。
如果要把一个字符串转换成 Uri，可以使用 Uri 类中的 parse()方法。

Uri uri = Uri.parse("content://com.siit.dict.DictProvider/word")

因为 Uri 代表了要操作的数据，所以经常需要解析 Uri，并从 Uri 中获取数据。Android

系统提供了两个用于操作 Uri 的工具类，分别为 UriMatcher 和 ContentUris。掌握它们的使用，有利于开发工作。

● UriMatcher 类用于匹配 Uri，它的用法如下。

① 首先把需要匹配 Uri 路径全部给注册上。

常量 UriMatcher.NO_MATCH 表示不匹配任何路径的返回码。

UriMatcher sMatcher = new UriMatcher(UriMatcher.NO_MATCH);

如果 match()方法匹配 content://com.siit.dict.DictProvider/word 路径，返回匹配码为 1，sMatcher.addURI("com.siit.dict.DictProvider", "word", 1)。添加需要匹配 uri，如果匹配就会返回匹配码。

如果 match()方法匹配 content:// com.siit.dict.DictProvider/word/230 路径，返回匹配码为 2，sMatcher.addURI("com.siit.dict.DictProvider", "word /#", 2);//#号为通配符。

```
private static final int WORDS = 1;
private static final int WORDS_ID = 2;
private static HashMap<String, String> wordProjectionMap;
static {
    // Uri 匹配工具类
    sUriMatcher = new UriMatcher(UriMatcher.NO_MATCH);
    sUriMatcher.addURI(Words.AUTHORITY, "word1", WORDS);
    sUriMatcher.addURI(Words.AUTHORITY, "word1/#", WORDS_ID);
    // 实例化查询列集合
    wordProjectionMap = new HashMap<String, String>();
    // 添加查询列
    wordProjectionMap.put(Words._ID, Words._ID);
    wordProjectionMap.put(Words.WORD, Words.WORD);
    wordProjectionMap.put(Words.DETAIL, Words.DETAIL);
}
```

② 注册完需要匹配的 Uri 后，在数据的操作方法中（如 insert()、delete()、update()、query()）就可以使用 sMatcher.match(uri)方法对输入的 Uri 进行匹配，如果匹配就返回匹配码，匹配码是调用 addURI()方法传入的第三个参数，假设匹配 content://com.example. provider.DictProvider/word 路径，返回的匹配码为 1。

```
switch(sUriMatcher.match(uri)){
case WORDS:
    …
    break;
case WORDS_ID:
    …
    break;
```

● ContentUris 类用于获取 Uri 路径后面的 ID 部分，它有两个比较实用的方法。

① withAppendedId(uri, id)用于为路径加上 ID 部分：

Uri uri = Uri.parse("content://com.siit.dict.DictProvider/word")

Uri resultUri = ContentUris.withAppendedId(uri, 10);

生成后的 Uri 为：content://com.siit.dict.DictProvider/word/10")。

② parseId(uri)方法用于从路径中获取 ID 部分：

Uri uri = Uri.parse("content:// com.siit.dict.DictProvider/word/10")

long personid = ContentUris.parseId(uri);//获取的结果为:10。

3. ContentProvider 的定义

当应用需要通过 ContentProvider 对外共享数据时，第一步需要继承 ContentProvider 并重写下面方法。

```java
public class DictProvider extends ContentProvider{
public boolean onCreate()
public Uri insert(Uri uri, ContentValues values)
public int delete(Uri uri, String selection, String[] selectionArgs)
public int update(Uri uri, ContentValues values, String selection, String[] selectionArgs)
public Cursor query(Uri uri, String[] projection, String selection, String[] selectionArgs, String sortOrder)
public String getType(Uri uri)
}
```

下面列出删除方法 delete( )的实现代码。其他方法的实现思路也与删除方法类似。具体内容详见任务实现的代码。

```java
public int delete(Uri uri, String selection, String[] selectionArgs) {
    int count=0;
    SQLiteDatabase db = dbHelper.getWritableDatabase();
    // 获得数据库实例
switch(sUriMatcher.match(uri)){
case WORDS:
    count=db.delete(WordDBHelper.WORDS_TableTbl, selection, selectionArgs);
case WORDS_ID:
    //通过 Uri 传递参数 temp
    long wordid=ContentUris.parseId(uri);
    String whereClause=Words._ID+"="+wordid;
    count=db.delete(WordDBHelper.WORDS_TableTbl,whereClause, selectionArgs);
    break;
}
```

ContentProvider 类的主要方法及其作用如下。

- public boolean onCreate() 该方法在 ContentProvider 创建后就会被调用，Android 开机后 ContentProvider 在其他应用第一次访问时会被创建。
- public Uri insert(Uri uri, ContentValues values) 该方法用于供外部应用向 ContentProvider 添加数据。
- public int delete(Uri uri, String selection, String[] selectionArgs) 该方法用于供外部应用从 ContentProvider 删除数据。
- public int update(Uri uri, ContentValues values, String selection, String[] selectionArgs) 该方法用于供外部应用更新 ContentProvider 中的数据。
- public Cursor query(Uri uri, String[] projection, String selection, String[] selectionArgs, String sortOrder) 该方法用于供外部应用从 ContentProvider 中获取数据。
- public String getType(Uri uri) 该方法用于返回当前 Uri 所代表数据的 MIME 类型。

第二步需要在 AndroidManifest.xml 的< application >结点中使用<provider>对该 ContentProvider 进行配置。为了能让其他应用找到该 ContentProvider，ContentProvider 采用了 authorities（主机名/域名）对它进行唯一标识，可以把 ContentProvider 看作是一个网站，authorities 就是域名。

```
<provider
    android:name=" com.siit.dict.DictProvider"
    android:authorities="com.siit.dict.DictProvider"
    android:exported="true" />
```

4. ContentProvider 的使用

ContnetResolver 是用来操作 ContentProvider 中的数据，Context 提供 getContentResolver() 方法来获取 ContentResolver 对象。

getContentResolver()一旦在程序中获得 ContentResolver 对象之后，接下来就可以调用 ContentResolver 如下方法来操作数据。

Insert(Uri uri，ContentValues values)：根据 Uri 插入 values 对应的数据。

Delete(Uri uri，String selection，String[ ] selectionArgs)：根据 Uri 删除 select 条件所匹配的全部记录。

Update(Uri uri，ContentValues values，String selection，String[ ] selectionArgs)：根据 Uri 修改 select 条件所匹配的全部记录。

Query(Uri uri，String[ ] projection,String selection，String[ ] selectionArgs，String sortOrder)：根据 Uri 查询出 select 条件所匹配的全部记录，其中 projection 就是一个列名列表，表明只选择出指定的数据列。

5. AndroidManifest 文件中的< provider>元素

Content Provider 是 ContentProvider 的子类，它为应用程序管理的数据提供结构化的访问方式。应用程序的所有 Content Provider 都必须在 AndroidManifest.xml 文件的< provider >元素中定义，否则系统不会运行 Content Provider。下面介绍< provider >元素的内容。

android:name 属性指向 Provider 类的定义包名和类名，在本任务中的 DictProvider 所属包名为 com.siit.dict，因此声明 android:name="com.siit.dict. DictProvider "。

android:authorities 属性表示一个或多个 URI authority 列表，标识了 Content Provider 内提供的数据。多个 authority 名称之间用分号分隔。为了避免冲突 authority 名称应该使用 Java 风格的命名规则（比如 com.siit.dict. DictProvider）。一般来说，它是实现 Content Provider 的 ContentProvider 子类名，没有默认值至少必须指定一个 authority。只需要声明本应用程序所含的 Content Provider 即可。属于其他应用程序而本程序会用到的 Content Provider 不需要进行声明。

android:exported 表示本 Content Provider 能否被其他应用程序使用，"true"为可以，任何应用程序都可以通过 URI 访问本 Content Provider，且受限于 Content Provider 声明的权限要求；"false"为不可以，通过设置 android:exported="false"，可以限制对本应用程序中的 Content Provider 进行。只有用户 ID 相同的应用程序才能访问到它。

任务分析

1．定义 ContentProvider

在 dict 工程中定义 ContentProvider，将 dict 工程中的内容共享给其他应用程序访问。

① 定义常量类 Words。
② 定义继承自 ContentProvider 的 DictProvider 类。
③ 在 AndroidManifest.xml 中声明 DictProvider。

2．实现不同应用间的内容共享

创建 CPTest 工程，在该工程中引用 Dict 的 ContentProvider 实现不同应用间的内容共享。

① 定义常量类 Words。
② 定义 ContentResolver 实现单词查询。

任务实现

1．定义 ContentProvider

（1）定义常量类 Words

定义常量类 Words，该类主要是用来声明一些常量，如表名、列名、排序 URI 和内容类型等。

```
import android.net.Uri;
import android.provider.BaseColumns;
public final class Words
{
    // 授权常量
    public static final String AUTHORITY = "com.siit.dict.DictProvider";
    // 定义一个静态内部类，定义该 ContentProvider 所包的数据列的列名
```

```java
public static final class Word implements BaseColumns
{
    public static final String WORDS_TableTbl = "wordtable";
    // 定义Content所允许操作的3个数据列
    public final static String _ID = "_id";
    public final static String WORD = "word";
    public final static String DETAIL = "detail";
    // 定义该Content提供服务的两个Uri
    public final static Uri DICT_CONTENT_URI = Uri
            .parse("content://" + AUTHORITY + "/word1");
}
}
```

（2）定义继承自 ContentProvider 的 DictProvider 类

定义 DictProvider 类继承自 ContentProvider，实现其抽象方法来操作数据库。该类通过创建 WordDBHelper 对象来获得数据库实例，引用 UriMatcher 工具类来对 URI 进行匹配比较，引用 HashMap 保存查询列，在 onCreate 方法中实例化 DBHelper 获得数据实例。用 insert()方法实现插入数据，delete()方法实现删除数据，query()方法实现查询数据，update()方法实现更新数据。

```java
import java.util.HashMap;
import android.content.ContentProvider;
import android.content.ContentUris;
import android.content.ContentValues;
import android.content.UriMatcher;
import android.database.Cursor;
import android.database.sqlite.SQLiteDatabase;
import android.net.Uri;
public class DictProvider extends ContentProvider
{
    // 数据库帮助类
    private WordDBHelper dbHelper;
    // Uri工具类
    private static final UriMatcher sUriMatcher;
    // 查询、更新条件
    private static final int WORDS = 1;
    private static final int WORDS_ID = 2;
    // 查询列集合
    private static HashMap<String, String> wordProjectionMap;
```

```java
static {
    // Uri 匹配工具类
    sUriMatcher = new UriMatcher(UriMatcher.NO_MATCH);
    sUriMatcher.addURI(Words.AUTHORITY, "word1", WORDS);
    sUriMatcher.addURI(Words.AUTHORITY, "word1/#", WORDS_ID);
    // 实例化查询列集合
    wordProjectionMap = new HashMap<String, String>();
    // 添加查询列
    wordProjectionMap.put(Words.Word._ID, Words.Word._ID);
    wordProjectionMap.put(Words.Word.WORD, Words.Word.WORD);
    wordProjectionMap.put(Words.Word.DETAIL, Words.Word.DETAIL);
}
@Override
public int delete(Uri uri, String selection, String[] selectionArgs) {
    int count=0;
    SQLiteDatabase db = dbHelper.getWritableDatabase();
    // 获得数据库实例
    switch(sUriMatcher.match(uri)){
    case WORDS:
        count=db.delete(Words.Word.WORDS_TableTbl, selection, selectionArgs);
    case WORDS_ID:
        //通过 Uri 传递参数 temp
        long wordid=ContentUris.parseId(uri);
        String whereClause=Words.Word._ID+"="+wordid;
        count=db.delete(Words.Word.WORDS_TableTbl,whereClause, selectionArgs);
        break;
    }
    return count;
}
@Override
public String getType(Uri uri) {
    switch (sUriMatcher.match(uri))
    {
    // 如果操作的数据是多项记录
    case WORDS:
        return "com.example.word.WordProvider/word1";
    // 如果操作的数据是单项记录
    case WORDS_ID:
```

```java
            return "com.example.word.WordProvider/word1/#";
        default:
            throw new IllegalArgumentException("未知Uri:" + uri);
    }
}
@Override
public Uri insert(Uri uri, ContentValues values) {
    // 获得数据库实例
    SQLiteDatabase db = dbHelper.getReadableDatabase();
    switch (sUriMatcher.match(uri))
    {
        // 如果Uri参数代表操作全部数据项
        case WORDS:
            // 插入数据，返回插入记录的ID
            long rowId=db.insert(Words.Word.WORDS_TableTbl, Words.Word._ID, values);
            // 如果插入成功返回uri
            if (rowId > 0)
            {
                // 在已有的Uri的后面追加ID
                Uri wordUri = ContentUris.withAppendedId(uri, rowId);
                // 通知数据已经改变
                getContext().getContentResolver().notifyChange(wordUri, null);
                return wordUri;
            }
            break;
        default :
            throw new IllegalArgumentException("未知Uri:" + uri);
    }
    return null;
}
@Override
public boolean onCreate() {
    // 实例化数据库帮助类
    dbHelper = new WordDBHelper(getContext());
    return true;
}
@Override
public Cursor query(Uri uri, String[] projection, String where,
```

```java
                String[] whereArgs, String sortOrder)
        {
            SQLiteDatabase db = dbHelper.getReadableDatabase();
            switch (sUriMatcher.match(uri))
            {
                // 如果Uri参数代表操作全部数据项
                case WORDS:
                    // 执行查询
                    return db.query(Words.Word.WORDS_TableTbl, projection, where,
whereArgs, null, null, sortOrder);
                // 如果Uri参数代表操作指定数据项
                case WORDS_ID:
                    // 解析出想查询的记录ID
                    long id = ContentUris.parseId(uri);
                    String whereClause = Words.Word._ID + "=" + id;
                    // 如果原来的where子句存在，拼接where子句
                    if (where != null && !"".equals(where))
                    {
                        whereClause = whereClause + " and " + where;
                    }
                    return db.query(Words.Word.WORDS_TableTbl, projection, whereClause,
whereArgs, null, null, sortOrder);
                default:
                    throw new IllegalArgumentException("未知Uri:" + uri);
            }
        }
        @Override
        public int update(Uri uri, ContentValues values, String where, String[]
whereArgs) {
            SQLiteDatabase db = dbHelper.getWritableDatabase();
            // 记录所修改的记录数
            int num = 0;
            switch (sUriMatcher.match(uri))
            {
                // 如果Uri参数代表操作全部数据项
                case WORDS:
                    num = db.update(Words.Word.WORDS_TableTbl, values, where,
whereArgs);
```

```
                break;
            // 如果 Uri 参数代表操作指定数据项
        case WORDS_ID:
            // 解析出想修改的记录 ID
            long id = ContentUris.parseId(uri);
            String whereClause = Words.Word._ID + "=" + id;
            // 如果原来的 where 子句存在，拼接 where 子句
            if (where != null && !where.equals(""))
            {
                whereClause = whereClause + " and " + where;
            }
            num = db.update(Words.Word.WORDS_TableTbl, values, whereClause, whereArgs);
            break;
        default:
            throw new IllegalArgumentException("未知Uri:" + uri);
        }
        // 通知数据已经改变
        getContext().getContentResolver().notifyChange(uri, null);
        return num;
    }
}
```

（3）在 AndroidManifest.xml 中声明 DictProvider

在 AndroidManifest.xml 文件定义< provider >元素，内容如下。

```
<!-- 注册一个 ContentProvider -->
        <provider
            android:name="com.siit.dict. DictProvider "
            android:authorities="com.siit.dict.DictProvider"
            android:exported="true" />
```

2．实现不同应用间的内容共享

创建 CPTest 工程，在该工程中通过 ContentProvider 获取 dict 工程的内容，并在 CPTest 工程中实现对 dict 当中单词信息的查询。

（1）定义常量类 Words

定义常量类 Words，该类主要是用来声明一些常量，如表名、列名、排序 URI 和内容类型等。

该类中声明 AUTHORITY = com.example. provider. DictProvider 是用于 ContentResolver 中的 URI 地址设置。注意必须 AndroidManifest.xml 中声明的一致。

```
import android.net.Uri;
```

```java
import android.provider.BaseColumns;
public final class Words
{
    // 授权常量
    public static final String AUTHORITY = " com.siit.dict.DictProvider";
    // 定义一个静态内部类，定义该ContentProvider所包的数据列的列名
    public static final class Word implements BaseColumns
    {
        public static final String WORDS_TableTbl = "wordtable";
        // 定义Content所允许操作的3个数据列
        public final static String _ID = "_id";
        public final static String WORD = "word";
        public final static String DETAIL = "detail";
        // 定义该Content提供服务的两个Uri
        public final static Uri DICT_CONTENT_URI = Uri
                .parse("content://" + AUTHORITY + "/word1");
    }
}
```

（2）创建 ContentResolver 实现单词的查询

创建 WordShowActivity，通过 ContentResolver 调用 DictProvider 的单词内容，实现对单词查询功能，通过 ContentResolver 调用 query 实现单词查看功能。

```java
package com.example.CPTest;
import android.app.Activity;
import android.app.AlertDialog;
import android.content.ContentResolver;
import android.content.Intent;
import android.database.Cursor;
import android.net.Uri;
import android.os.Bundle;
import android.provider.ContactsContract.Contacts;
import android.view.View;
import android.view.View.OnClickListener;
import android.widget.Button;
import android.widget.EditText;
import android.widget.ListView;
import android.widget.SimpleAdapter;
import android.widget.SimpleCursorAdapter;
import android.widget.Toast;
```

```java
public class WordShowActivity extends Activity {
    ContentResolver cr;
    private EditText word,detail;
    private Button btnprior,btnnext,btnedit;
    Cursor c ;
    @Override
    protected void onCreate(Bundle savedInstanceState) {
        // TODO Auto-generated method stub
        super.onCreate(savedInstanceState);
        setContentView(R.layout.wordshow_activity);
        word=(EditText) findViewById(R.id.word);
        detail=(EditText) findViewById(R.id.detail);
        btnprior=(Button) findViewById(R.id.btnprior);
        btnnext=(Button) findViewById(R.id.btnnext);
        btnedit=(Button) findViewById(R.id.editbtn);
        cr=getContentResolver();
        c=cr.query(Words.Word.CONTENT_URI, null, null, null, null);
        if(c.getCount()>0){
            c.moveToFirst();
            word.setText(c.getString(1));
            detail.setText(c.getString(2));
        }
        else
            Toast.makeText(getApplicationContext(), "没有单词了！", 3000).show();
        btnprior.setOnClickListener(new OnClickListener() {
            @Override
            public void onClick(View arg0) {
                if(c.moveToPrevious()){
                    word.setText(c.getString(1));
                    detail.setText(c.getString(2));
                }
                else
                {
                    Toast.makeText(getApplicationContext(), "已经是第一条单词", 3000).show();
                }
            }
```

```
            });
            btnnext.setOnClickListener(new OnClickListener() {
                @Override
                public void onClick(View v) {
                    // TODO Auto-generated method stub
                    if(c.moveToNext()){
                        word.setText(c.getString(1));
                        detail.setText(c.getString(2));}
                    else
                    {
                        Toast.makeText(getApplicationContext(), "已经是最后一条单词", 3000).show();
                    }
                }
            });
        }
        @Override
        protected void onResume() {
            cr=getContentResolver();
            c=cr.query(Words.Word.CONTENT_URI, null, null, null, null);
            if(c.getCount()>0){
                c.moveToFirst();
                word.setText(c.getString(1));
                detail.setText(c.getString(2));
            }
            else
                Toast.makeText(getApplicationContext(), "没有单词了！", 3000).show();
            super.onResume();
        }
    }
```

拓展学习

使用 ContentProvider 实现通信录功能的设计与实现。通信录功能包括：编号、姓名、电话、电子邮件，设计应用程序实现通信录的添加、删除、修改和查询。

任务 2.7　用户信息网络传输

任务描述

使用 Apache Http 方法实现 Android 客户端和 Web 服务器端的数据交互，并通过 Android 客户端向 Web 服务器端发送数据，实现用户登录和用户注册的功能，如图 2-29 所示，数据添加成功后在 MySQL 中的数据显示如图 2-30 所示。

图 2-29　连接服务器的用户登录和注册运行效果

图 2-30　用户信息表数据

任务目标

① 理解 Apache Http 方式下 Android 客户端与 Web 服务器端的请求响应机制。
② 掌握 Android 客户端的网络请求响应处理。
③ 掌握 Web 服务器端利用 Servlet 实现请求响应处理的机制。

任务分析

1. Web 服务器端的操作

① 使用 MySQL 实现用户表的设计与创建。

② 连接数据库实现用户登录与注册的数据处理方法。
③ 服务器端登录与注册功能的 HttpServlet 设计与实现。

2．Android 客户端的操作

① 使用 Apache Http 方法连接网络。
② 实现用户登录和注册功能的请求。

知识要点

为了处理客户端向 Web 站点的请求，Apache 开源组织提供了一个 HttpClient 项目，该项目是一个简单的 HTTP 客户端，可以用于发送 HTTP 请求，接收 HTTP 响应。Android 已经集成了 HttpClient，开发人员可以直接在 Android 中使用 HttpClient 来访问提交请求、接收响应。下面介绍 Apache HttpClient 实现网络数据请求与响应的相关内容。

1．使用 Apache HttpClient 访问网络的基本步骤

创建 HttpGet 或 HttpPost 对象，将要请求的 URL 通过构造方法传入 HttpGet 或 HttpPost 对象。

使用 DefaultHttpClient 类的 execute 方法发送 HTTP GET 或 HTTP POST 请求，并返回 HttpResponse 对象。

通过 HttpResponse 接口的 getEntity 方法返回响应信息，并进行相应的处理。如果使用 HttpPost 方法提交 HTTP POST 请求，还需要使用 HttpPost 类的 setEntity 方法设置请求参数。

Web 服务器和 Android 客户端之间网络数据传输过程如图 2-31 所示。

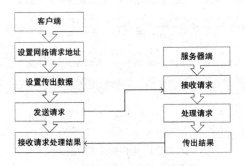

图 2-31　网络请求响应的步骤

2．HTTP 中 Get 与 Post 方法的区别

表单提交中 Get 和 Post 方法的区别有以下 5 点。

① Get 是从服务器上获取数据，Post 是向服务器传送数据。

② Get 是把参数数据队列加到提交表单的 ACTION 属性所指的 URL 中，值和表单内各个字段一一对应，在 URL 中可以看到。Post 是通过 HTTP post 机制，将表单内各个字段与其内容放置在 HTML HEADER 内一起传送到 ACTION 属性所指的 URL 地址。用户看不到这个过程。

③ 对于 Get 方式，服务器端用 Request.QueryString 获取变量的值；对于 Post 方式，服务器端用 Request.Form 获取提交的数据。

④ Get 传送的数据量较小，不能大于 2KB；Post 传送的数据量较大，一般被默认为不受限制。但理论上，IIS4 中最大量为 80KB，IIS5 中为 100KB。

⑤ Get 安全性非常低，Post 安全性较高。

3. Android 客户端数据请求与响应

（1）定义网络请求及请求地址

```
String urlStr = "http://192.168.0.100:8080/UserManageSystem/LoginServlet";
HttpPost request = new HttpPost(urlStr);
```

（2）设置传出数据

```
List<NameValuePair> params = new ArrayList<NameValuePair>();
params.add(new BasicNameValuePair("id", id));
request.setEntity( new UrlEncodedFormEntity(params,HTTP.UTF_8));
```

（3）执行请求

```
HttpResponse response = new DefaultHttpClient().execute(request);
```

（4）接收请求在服务器端的处理结果

```
if(response.getStatusLine().getStatusCode()==200){
        String msg = EntityUtils.toString(response.getEntity());
        showDialog(msg);
}
```

4. Web 服务器端请求与响应

（1）接收请求

定义 Servlet 接收请求，重写 doGet()与 doPost()方法。

```
public class DeleteUserServlet extends HttpServlet{}
```

（2）处理请求，传出处理结果

```
PrintWriter out = response.getWriter();
int userid = Integer.parseInt(request.getParameter("userid"));
String msg = null;
UserDao dao = new UserDaoImpl();
int row = dao.deleteUser(userid);
if(row>=0)
msg="用户删除成功！";
else
msg="用户删除不成功！";
out.print(msg);
```

（3）配置 Servlet

```
<servlet>
<servlet-name>DeleteUserServlet</servlet-name>
<servlet-class>com.usermanagesystem.http.DeleteUserServlet</servlet-class>
```

```xml
</servlet>
<servlet-mapping>
<servlet-name>DeleteUserServlet</servlet-name>
<url-pattern>/servlet/DeleteUserServlet</url-pattern>
</servlet-mapping>
```

任务实现

1. Web 服务器端的操作

(1) 使用 MySQL 实现用户表的设计与创建

```sql
CREATE TABLE 'usertbl' (
  'id' int(11) NOT NULL AUTO_INCREMENT COMMENT '主键,自动增加',
  'account' varchar(20) DEFAULT NULL COMMENT '登录账号',
  'password' varchar(20) DEFAULT NULL COMMENT '登录密码',
  'name' varchar(20) DEFAULT NULL COMMENT '姓名',
  'gender' varchar(20) DEFAULT NULL COMMENT '性别',
  'favorite' varchar(100) DEFAULT NULL COMMENT '权限 1:管理员 2:收银员 3:服务员',
  'remark' varchar(200) DEFAULT NULL COMMENT '备注',
  PRIMARY KEY ('id')
) ENGINE=InnoDB AUTO_INCREMENT=12 DEFAULT CHARSET=utf8;
```

(2) 连接数据库实现用户登录与注册的方法

网络服务器端用户登录与注册功能实现，用户登录与注册功能基于 Java 的 MVC 模式，下面介绍服务器端连接数据库的实现代码。数据库连接工具类 DBUtil、用户实体类为 User 类、用户数据操作接口 UserDao、用户数据操纵的实现类 UserDaoImpl。处理用户登录的 LoginServlet 类和用户注册的 InsertUserServlet 类。

数据库连接工具 DBUtil 内容如下。

```java
public class DBUtil {
    /**
     * 关闭数据库连接
     */
    public void closeConn(Connection conn){
        try {
            conn.close();
        } catch (SQLException e) {
            e.printStackTrace();
        }
    }
    /**
```

```
 * 打开数据库连接
 */
public Connection openConnection() {
    Properties prop = new Properties();
    String driver = null;
    String url = null;
    String username = null;
    String password = null;
    try {
        prop.load(this.getClass().getClassLoader().getResourceAsStream(
        "DBConfig.properties"));
        driver = prop.getProperty("driver");
        url = prop.getProperty("url");
        username = prop.getProperty("username");
        password = prop.getProperty("password");
        Class.forName(driver);
        return DriverManager.getConnection(url, username, password);
    } catch (Exception e) {
        e.printStackTrace();
    }
    return null;
}
```

DBConfig.properties 文件内容如下。

```
driver=com.mysql.jdbc.Driver
url=jdbc:mysql://localhost:3306/word_db?useUnicode=true&characterEncoding=utf-8
username=root
password=1
```

用户实体类 User 内容如下。

```
public class User {
    // 编号
    private int id;
    // 账号
    private String account;
    // 密码
    private String password;
    // 用户名称
```

```java
        private String name;
        // 性别
        private String gender;
        // 权限
        private String favorite;
        // 备注
        private String remark;

        public String getAccount() {
            return account;
        }
        public void setAccount(String account) {
            this.account = account;
        }
        public String getGender() {
            return gender;
        }
        public void setGender(String gender) {
            this.gender = gender;
        }
        public int getId() {
            return id;
        }
        public void setId(int id) {
            this.id = id;
        }
        public String getName() {
            return name;
        }
        public void setName(String name) {
            this.name = name;
        }
        public String getPassword() {
            return password;
        }
        public void setPassword(String password) {
            this.password = password;
        }
```

```java
    public String getFavorite() {
        return favorite;
    }
    public void setFavorite(String favorite) {
        this.favorite = favorite;
    }
    public String getRemark() {
        return remark;
    }
    public void setRemark(String remark) {
        this.remark = remark;
    }
}
```

用户数据操作接口 UserDao 内容如下。

```java
public interface UserDao {
    // 登录方法
    User login(String account,String password);
    public int insertuser(User u);
    public User findUserByid(int userid);
    public int updateUser(User u);
    publici nt deleteUser(int userid);
    public List<User> userList();
}
```

用户数据操纵的实现类 UserDaoImpl 内容如下。

```java
public class UserDaoImpl implements UserDao {
    private String sql;
    PreparedStatement psmt;
    DBUtil util;
    Connection conn;
    /**
     *通过用户名称和密码登录，登录成功返回 User 对象，登录失败返回 null
     */
    public User login(String account, String password) {
        // 查询 SQL 语句
        sql = " select id,account,password,name,favorite,remark "+
        " from userTbl "+
        " where account=? and password=? ";
        try {
```

```java
            // 获得预定义语句
            psmt = conn.prepareStatement(sql);
            // 设置查询参数
            psmt.setString(1, account);
            psmt.setString(2, password);
            // 执行查询
            ResultSet rs = psmt.executeQuery();
            // 判断用户是否存在
            if (rs.next()) {
                // 获得用户信息
                int id = rs.getInt(1);
                String name = rs.getString(4);
                String favorite = rs.getString(5);
                String remark = rs.getString(6);
                // 封装用户信息
                User u = new User();
                u.setId(id);
                u.setAccount(account);
                u.setPassword(password);
                u.setName(name);
                u.setFavorite(favorite);
                u.setRemark(remark);
                return u;
            }
        } catch (SQLException e) {
            e.printStackTrace();
        }
        return null;
    }
    public int insertuser(User u) {
        int row = 0 ;
        sql = "insert into userTbl(account,password,name,gender,favorite) values(?,?,?,?,?)";
        try {
            psmt =conn.prepareStatement(sql);
            psmt.setString(1, u.getAccount());
            psmt.setString(2, u.getPassword());
            psmt.setString(3, u.getName());
```

```java
            psmt.setString(4,u.getGender());
            psmt.setString(5,u.getFavorite());
            row = psmt.executeUpdate();
        } catch (SQLException e) {
            System.out.print("添加用户信息: ");
            e.printStackTrace();
        }
        return row;
    }
    public UserDaoImpl() {
        super();
        // 数据库连接工具类
        util = new DBUtil();
        // 获得连接
        conn = util.openConnection();
    }
    public User findUserByid(int userid) {
        // 查询SQL语句
        sql = " select id,account,password,name,favorite,remark "+
              " from userTbl "+
              " where id=? ";
        try {
            // 获得预定义语句
            psmt = conn.prepareStatement(sql);
            // 设置查询参数
            psmt.setInt(1, userid);
            // 执行查询
            ResultSet rs = psmt.executeQuery();
            // 判断用户是否存在
            if (rs.next()) {
                // 获得用户信息
                int id = rs.getInt(1);
                String account=rs.getString(2);
                String password=rs.getString(3);
                String name = rs.getString(4);
                String favorite = rs.getString(5);
                String remark = rs.getString(6);
                // 封装用户信息
```

```java
            User u = new User();
            u.setId(id);
            u.setAccount(account);
            u.setPassword(password);
            u.setName(name);
            u.setFavorite(favorite);
            u.setRemark(remark);
            return u;
        }
    } catch (SQLException e) {
        e.printStackTrace();
    }
    return null;
}
public int deleteUser(int userid) {
    int row=0;
    // 查询SQL语句
    sql = " delete userTbl where id=?  ";
    try {
        // 获得预定义语句
        psmt = conn.prepareStatement(sql);
        // 设置查询参数
        psmt.setInt(1, userid);
        // 执行查询
        row=psmt.executeUpdate();
    } catch (SQLException e) {
        e.printStackTrace();
    }
    return row;
}
public int updateUser(User u) {
    int row = 0 ;
    sql = "update userTbl set account=?,password=?,name=?,gender=?,favorite=? where id=?";
    try {
        psmt =conn.prepareStatement(sql);
        psmt.setString(1,u.getAccount());
        psmt.setString(2,u.getPassword());
```

```java
            psmt.setString(3,u.getName());
            psmt.setString(4,u.getGender());
            psmt.setString(5,u.getFavorite());
            psmt.setInt(6,u.getId());
            row = psmt.executeUpdate();
        } catch (SQLException e) {
            System.out.print("添修改用户信息：");
            e.printStackTrace();
        }
        return row;
    }
    public List<User> userList() {
        List<User> users=new ArrayList<User>();
        sql="select id,account,password,name,gender,favorite,remark "+
        " from userTbl ";
        try {
            psmt = conn.prepareStatement(sql);
            ResultSet rs=psmt.executeQuery();
            while(rs.next()){
                User u = new User();
                u.setId(rs.getInt("id"));
                u.setAccount(rs.getString("account"));
                u.setGender(rs.getString("gender"));
                u.setPassword(rs.getString("password"));
                u.setName(rs.getString("name"));
                u.setFavorite(rs.getString("favorite"));
                u.setRemark(rs.getString("remark"));
                users.add(u);}
        } catch (SQLException e) {
            // TODO Auto-generated catch block
            e.printStackTrace();
        }
        return users;
    }
}
```

（3）服务器端登录与注册功能的 HttpServlet 设计与实现

创建 LoginServlet 实现用户登录信息验证。该类继承自 HttpServlet，接收 Android 客户端发送过来的用户名和密码，结合 Web 服务器端的数据处理类实现用户名和密码的验证。如果数据库中存在一条记录和 Android 发送过来的信息一致，则返回用户信息，否则返回 0。

LoginServlet 功能代码如下。

```java
public class LoginServlet extends HttpServlet {
    public void doGet(HttpServletRequest request, HttpServletResponse response)
            throws ServletException, IOException {
        String username = request.getParameter("username");
        String password = request.getParameter("password");
        System.out.println(username+":"+password);
        response.setContentType("text/html");
        response.setCharacterEncoding("utf-8");
        PrintWriter out = response.getWriter();
        String msg = null;
        UserDao dao = new UserDaoImpl();
        User u = dao.login(username, password);
        if(u!=null){
            // 响应客户端内容，登录成功
            out.print(u.getId()+","+u.getName()+","+u.getGender()+","+u.getFavorite());
        }
        else{
            out.print("0");
        }
        out.flush();
        out.close();
    }

    publicvoid doPost(HttpServletRequest request, HttpServletResponse response)
            throws ServletException, IOException {
        doGet(request,response);
    }
}
```

用户注册 Servlet，接收从 Android 端发送过来的用户名、密码、姓名、性别、爱好等信息，并将其添加到数据库中，添加成功返回"用户添加成功"，否则返回"用户添加失败"。

```java
public class InsertUserServlet extends HttpServlet {
    public void doGet(HttpServletRequest request, HttpServletResponse
```

```java
response)
        throws ServletException, IOException {
        String account = request.getParameter("account");
        String password = request.getParameter("password");
        String name = request.getParameter("name");
        String gender = new String(request.getParameter("gender").getBytes("iso-8859-1"),"utf-8");
        //String question = new String(request.getParameter ("question").getBytes("iso-8859-1"),"utf-8");
        String favorite = request.getParameter("favorite");
        System.out.println(account+":"+password+":"+name+":"+gender+":"+favorite);
        response.setContentType("text/html");
        response.setCharacterEncoding("utf-8");
        PrintWriter out = response.getWriter();
        System.out.println("添加用户信息：");
        String msg = null;
        UserDao dao = new UserDaoImpl();
        User user = new User();
        user.setAccount(account);
        user.setPassword(password);
        user.setName(name);
        user.setGender(gender);
        user.setFavorite(favorite);
            int i=dao.insertuser(user);
        if(i>=0)
            msg="用户添加成功！";
        else
            msg="用户添加不成功！";
        out.print(msg);
        System.out.print(msg);
        out.flush();
        out.close();
    }
    public void doPost(HttpServletRequest request, HttpServletResponse response)
        throws ServletException, IOException {
        doGet(request,response);
```

            }
    }

登录和注册功能的 web.Xml 中 Sevlet 配置如下。

```xml
<servlet>
<servlet-name>LoginServlet</servlet-name>
    <servlet-class>com.usermanagesystem.http.LoginServlet</servlet-class>
</servlet>
<servlet>
<servlet-name>InsertUserServlet</servlet-name>
    <servlet-class>com.usermanagesystem.http.InsertUserServlet</servlet-class>
</servlet>
<servlet-mapping>
<servlet-name>LoginServlet</servlet-name>
<url-pattern>/LoginServlet</url-pattern>
</servlet-mapping>
<servlet-mapping>
<servlet-name>InsertUserServlet</servlet-name>
<url-pattern>/InsertUserServlet</url-pattern>
</servlet-mapping>
```

2．Android 客户端的操作

① 在 Android 客户端功能实现用户登录与用户注册，此处主要介绍用户注册和用户登录的方法。

用户登录：从界面获取用户名、密码，通过 http 发送到服务器端。

```java
private String login(String username,String password){      // IP 地址需要根据实际情况设置
        String returnstr="";
        String urlStr = "http://192.168.0.100:8080/UserManageSystem/LoginServlet";
        HttpPost request = new HttpPost(urlStr);
        // 如果传递参数个数比较多的话，我们可以对传递的参数进行封装
        List<NameValuePair> params = new ArrayList<NameValuePair>();
        params.add(new BasicNameValuePair("username", username));
        params.add(new BasicNameValuePair("password", password));
        try {
            request.setEntity(new UrlEncodedFormEntity(params,HTTP.UTF_8));
            HttpResponse response = new DefaultHttpClient().execute(request);
            if(response.getStatusLine().getStatusCode()==200){
```

```
            returnstr = EntityUtils.toString(response.getEntity());
        }
    } catch (Exception e) {
        e.printStackTrace();
    }
    return returnstr;
}
```

② 实现用户注册功能,从界面获取用户名、密码、姓名、性别、爱好,通过 HttpClient 发送请求,代码如下。

```
private void registerUser(String username,String password,String name,
String gender,String favorite){
    // 2. 使用Apache HTTP 客户端实现
    String urlStr ="http://192.168.0.100:8080/UserManageSystem /InsertUserServlet";
    HttpPost request = new HttpPost(urlStr);
    // 如果传递参数个数比较多的话,我们可以对传递的参数进行封装
    List<NameValuePair> params = new ArrayList<NameValuePair>();
    params.add(new BasicNameValuePair("account", username));
    params.add(new BasicNameValuePair("password", password));
    params.add(new BasicNameValuePair("name", name));
    params.add(new BasicNameValuePair("gender", gender));
    params.add(new BasicNameValuePair("favorite",favorite));
    try {
        //request.addHeader("Content-Type", "text/html");    //这行很重要
        request.setEntity( new UrlEncodedFormEntity(params,HTTP.UTF_8));
        HttpResponse response = new DefaultHttpClient().execute(request);
        if(response.getStatusLine().getStatusCode()==200){
            String msg = EntityUtils.toString(response.getEntity());
            showDialog(msg);
        }
    } catch (Exception e) {
        e.printStackTrace();
    }
}
```

技能训练

在单词管理 WordManageActivity 类中完善单词管理功能,在原有 SQLite 数据库单词管理

的基础上，利用网络 Apache Http 方法实现将 Android 客户端的单词添加和单词更新到 Web 服务器端的 MySQL 数据库中，使得 SQLite 数据库与服务器端数据库同步，界面效果如图 2-32 所示，Web 服务器端的数据如图 2-34 所示。

图 2-32　运行效果图

任务 2.8　单词网络下载

任务描述

分别使用 JSON 和 XML 文件将 Web 服务器端数据库中的单词信息通过移动互联网传输至 Android 客户端，在 Android 客户端利用 JSON 和 XML 文件方式解析服务器端单词信息，并使用列表将单词信息显示到界面当中。

JSON 指的是 JavaScript 对象表示法（JavaScript Object Notation）。

XML 的全称为 Extensible Markup Language 可扩展标记语言。

界面和服务器端数据如图 2-33 和图 2-34 所示。

图 2-33　在 Android 端显示的单词结果图

图 2-34 Web 服务器端数据库

任务目标

① 掌握网络数据传输的过程和实现方法。
② 掌握利用 JSON 获取服务器端数据，解析 JSON 数据，利用列表展示 JSON 数据的方法。
③ 掌握利用 XML 获取服务器端数据，解析 XML 数据，利用列表展示 XML 数据的方法。

任务分析

该任务主要实现 Web 服务器端向 Andorid 客户端批量传输数据。在 Android 客户端使用 JSON 格式或者 XML 形式将单词信息进行解析和显示，具体实现过程如下。

1．Web 服务器端的操作

① 在 Web 端利用 MVC 模式实现单词信息查询。
② 在 Web 端单词将信息转换为 JSON 格式或转换为 XML 格式。

2．Android 客户端的操作

① 利用 URL 访问 Android 互联网。
② 客户端利用 JSON 或 XML 获取并解析单词信息。
③ 将其显示到界面当中。

知识要点

1．JSON 简介

JSON 指的是 JavaScript 对象表示法（JavaScript Object Notation），是轻量级的文本数据交换格式，具有自我描述性，独立于语言且更易理解。JSON 解析器和 JSON 库支持许多不同的编程语言。

JSPON 的基本格式：[{name:value, {name:value, {name:value,…},{{name:value, {name:value, {name:value,…},…]

在 Java 中解析 JSON 数据，首先需要加载 org.json 的 jar 包，然后使用 JSONArray 获取 JSON 数组，并利用 JSONObject 依次遍历 JSON 的每一个对象。解析 JSON 的核心语句如下，具体方法参见 parseJSON()方法。

生成 JSON 数组：JSONArray array=new JSONArray(json);
生成 JSON 对象：JSONObject jsonObject=array.getJSONObject(i);

2．XML 简介

XML 的全称为 Extensible Markup Language 可扩展标记语言，标准通用标记语言的子集，是一种用于标记电子文件使其具有结构性的标记语言。XML 文件示例：

```
<?xml version="1.0" encoding="UTF-8"?>
<student>
<sno>110011</sno>
    <name>George</name>
    <sex>男</sex>
</student>
```

能够运用在 Android 系统上解析 XML 文件的有 3 种常用方式：DOM、SAX 和 PULL，其中 DOM 解析 XML 是先把 XML 文件读进内存中，再通过接口获取数据，该方法使用相对小的 XML 文件，移动设备往往受硬件性能影响，如果 XML 文件比较大使用 DOM 解析往往效率跟不上；SAX 和 PULL 都是采用事件驱动方式来进行解析，在 Android 中的事件机制是基于回调函数。

此任务中使用的是 DOM 解析方式，XML 文件解析具体思路是：首先利用 DocumentBuilderFactory 创建一个 DocumentBuilderFactory 实例，再利用 DocumentBuilderFactory 创建 DocumentBuilder，加载 XML 文档（Document），获取文档的根结点（Element），然后获取根结点中所有子节点的列表（NodeList），使用再获取子节点列表中的需要读取的结点。XML 在 Android 的解析参见文中的 parseXML（InputStream inStream）方法。

任务实现

1. Web 服务器端的操作

（1）在 Web 端利用 MVC 模式实现单词信息查询

创建单词信息实体类、数据处理类和数据传输 Servlet。

```
public class Word {
    private int id;
    private String word;
    private String detail;
    public Word(){}
    public Word(int id, String word, String detail) {
        super();
        this.id = id;
        this.word = word;
        this.detail = detail;
    }
    public int getId() {
        return id;
    }
    public void setId(int id) {
```

```java
            this.id = id;
        }
        public String getWord() {
            return word;
        }
        public void setWord(String word) {
            this.word = word;
        }
        public String getDetail() {
            return detail;
        }
        public void setDetail(String detail) {
            this.detail = detail;
        }
}

public interface WordDao {
        public int insertWord(Word u);
        public Word findWordByid(int Wordid);
        public int updateWord(Word u);
        public int deleteWord(int Wordid);
        public List<Word> wordList();
}

public class WordDaoImpl implements WordDao {
        private String sql;
        PreparedStatement psmt;
        DBUtil util;
        Connection conn;
        public WordDaoImpl() {
            super();
            // 数据库连接工具类
            util = new DBUtil();
            // 获得连接
            conn = util.openConnection();
        }
        public List<Word> wordList() {
            List<Word> words=new ArrayList<Word>();
```

```java
            sql="select id,word,detail from wordTbl ";
            try {
                psmt = conn.prepareStatement(sql);
                ResultSet rs=psmt.executeQuery();
                while(rs.next()){
                    Word w = new Word();
                    w.setId(rs.getInt("id"));
                    w.setWord(rs.getString("word"));
                    w.setDetail(rs.getString("detail"));
                    words.add(w);}
        } catch (SQLException e) {
                // TODO Auto-generated catch block
                e.printStackTrace();
            }
            return words;
    }

    public int deleteWord(int Wordid) {
        int row=0;
        // 查询 SQL 语句
        sql = " delete wordTbl where id=?  ";
        try {
            // 获得预定义语句
            psmt = conn.prepareStatement(sql);
            // 设置查询参数
            psmt.setInt(1, Wordid);
            // 执行查询
            row=psmt.executeUpdate();
        } catch (SQLException e) {
            e.printStackTrace();
        }
        return row;
    }

    public Word findWordByid(int Wordid) {
        // 查询 SQL 语句
        sql = " select id,word,detail "+
        " from wordTbl "+
```

```java
            " where id=? ";
        try {
            // 获得预定义语句
            psmt = conn.prepareStatement(sql);
            // 设置查询参数
            psmt.setInt(1, Wordid);
            // 执行查询
            ResultSet rs = psmt.executeQuery();
            // 判断用户是否存在
            if (rs.next()) {
                // 获得用户信息
                int id = rs.getInt(1);
                String word=rs.getString(2);
                String detail=rs.getString(3);
                // 封装用户信息
                Word w = new Word();
                w.setId(id);
                w.setWord(word);
                w.setDetail(detail);
                return w;
            }
        } catch (SQLException e) {
            e.printStackTrace();
        }
        return null;
    }
    public int insertWord(Word w) {
        int row = 0 ;
        sql = "insert into wordTbl(word,detail) values(?,?)";
        try {
            psmt =conn.prepareStatement(sql);
            psmt.setString(1, w.getWord());
            psmt.setString(2, w.getDetail());
            row = psmt.executeUpdate();
        } catch (SQLException e) {
            System.out.print("添加单词信息：");
            e.printStackTrace();
        }
```

```java
            return row;
        }
        public int updateWord(Word w) {
            int row = 0 ;
            sql = "update wordTbl set word=?,detail=? where id=?";
            try {
                psmt =conn.prepareStatement(sql);
                psmt.setString(1, w.getWord());
                psmt.setString(2, w.getDetail());
                psmt.setInt(3, w.getId());
                row = psmt.executeUpdate();
            } catch (SQLException e) {
                System.out.print("添修改单词信息: ");
                e.printStackTrace();
            }
            return row;
        }
}
```

在 WordDaoImpl 中用 List<Word> wordList()方法以列表形式输出用户信息。

```java
public List<Word> wordList() {
    List<Word> words=new ArrayList<Word>();
    sql="select id,word,detail from wordTbl ";
    try {
        psmt = conn.prepareStatement(sql);
        ResultSet rs=psmt.executeQuery();
        while(rs.next()){
            Word w = new Word();
            w.setId(rs.getInt("id"));
            w.setWord(rs.getString("word"));
            w.setDetail(rs.getString("detail"));
            words.add(w);}
    } catch (SQLException e) {
        // TODO Auto-generated catch block
        e.printStackTrace();
    }
    return words;
}
```

（2）在 Web 端将单词信息转换为 JSON 格式或 XML 格式

定义 ListWordServlet 将用户信息以 JSON 格式或 XML 格式输出。

```java
public class ListWordServlet extends HttpServlet {
    public void doGet(HttpServletRequest request, HttpServletResponse response)
    throws ServletException, IOException {
        doPost(request,response);
    }
    public void doPost(HttpServletRequest request, HttpServletResponse response)
    throws ServletException, IOException {
        response.setContentType("text/html");
        response.setCharacterEncoding("utf-8");
        PrintWriter out = response.getWriter();
        List<Word> words=new WordDaoImpl().wordList();
        StringBuilder builder=new StringBuilder();
        String format=request.getParameter("format");
        if(format.equals("json")){
            builder.append('[');
            for(Word word:words){
                builder.append("{");
                builder.append("id:").append(word.getId()).append(',');
                builder.append("word:").append(word.getWord()).append(',');
                builder.append("detail:").append(word.getDetail());
                builder.append("},");
            }
            builder.deleteCharAt(builder.length()-1);
            builder.append(']');
            out.print(builder.toString());
        }
        else if(format.equals("xml")){
            builder.append("<?xml version='1.0' encoding='UTF-8'?>");
            builder.append("<wordlist>");
            for(Word word:words){
                builder.append("<wordcontent>");
                builder.append("<id>");
                builder.append(word.getId());
                builder.append("</id>");
```

```
                builder.append("<word>");
                builder.append(word.getWord());
                builder.append("</word>");
                builder.append("<detail>");
                builder.append(word.getDetail());
                builder.append("</detail>");
                builder.append("</wordcontent>");
            }
            builder.append("</wordlist>");
            out.print(builder.toString());
        }
        System.out.println(builder.toString());
        out.flush();
        out.close();
    }
}
```

2．Android 客户端的操作

（1）利用 URL 访问 Android 互联网

在 WordShowActivity 类中添加 getLastJsonUser()方法。该方法的作用是向指定 IP 发送请求，并接收 Web 服务器端处理的数据结果。

```
public static List<Word> getLastJsonUser() throws Exception{
      String path="http://192.168.0.100:8080/UserManageSystem/ListWordServlet?format=json";
      URL url=new URL(path);
      URLConnection conn=url.openConnection();
      conn.setConnectTimeout(5000);
      conn.setDoOutput(true);
      conn.setDoInput(true);
      InputStream inStream=conn.getInputStream();
      return parseJSON(inStream);
}
```

（2）客户端利用 JSON 或 XML 获取并解析单词信息

在 Android 客户端获取服务器端的处理结果集，然后利用 JSON 解析数据，并使用 JSONArray 获取 JSON 数组，使用 JSONObject 依次遍历 JSON 的每一个对象。在 WordShowActivity 中添加 parseJSON()方法，主要代码如下。

```
private static List<Word> parseJSON(InputStream inStream) throws Exception {
      List<Word> words=new ArrayList<Word>();
      byte[] data=com.example.util.StreamTool.read(inStream);
```

```
            String json=new String(data);
            JSONArray array=new JSONArray(json);
            for(int i=0;i<array.length();i++){
                for(int i=0;i<list.getLength();i++){
                Element wordElement = (Element)list.item(i);
                Word word=new Word(Integer.parseInt(wordElement.getElementsByTagName
("id").item(0).getFirstChild().getNodeValue()),wordElement.getElementsByTag
Name("word").item(0).getFirstChild().getNodeValue(),wordElement.getElements
ByTagName("detail").item(0).getFirstChild().getNodeValue());
            words.add(word);
        }
        return words;
    }
```

在 WordShowActivity 中添加 parseXML()方法，解析 XML 文档内容，主要代码如下。

```
    private static List<Word> parseXML(InputStream inStream) throws
Exception {
        List<Word> words=new ArrayList<Word>();
        // step 1：获得 dom 解析器工厂（工作的作用是用于创建具体的解析器）
        DocumentBuilderFactory factory=DocumentBuilderFactory.newInstance();
        // step 2：获得具体的 dom 解析器
        DocumentBuilder builder=factory.newDocumentBuilder();
        // step 3：解析一个 xml 文档，获得 Document 对象（根结点）
        Document doc=builder.parse(inStream);
//step4:获取 xml 中的 wordcontent 元素
        NodeList list=doc.getElementsByTagName("wordcontent");
// step5:利用 for 循环获取每个 id 元素、word 元素、detail 元素的值，并通过 Word 构造方
法创建 Word 对象，一次添加到 words 列表当中
        for(int i=0;i<list.getLength();i++){
            Wordword=newWord (
  Integer.parseInt(doc.getElementsByTagName("id").
item(i).getFirstChild().getNodeValue()),doc.getElementsByTagName("word").it
em(i).getFirstChild().getNodeValue(),doc.getElementsByTagName("detail").ite
m(i).getFirstChild().getNodeValue());
            words.add(word);
        }
        return words;
    }
```

(3) 将单词信息显示到界面当中

在 Android 客户端创建 Word 实体类。

```java
package com.example.remeberword;
public class Word {
    private int id;
    private String word;
    private String detail;
    public Word(){}
    public Word(int id, String word, String detail) {
        super();
        this.id = id;
        this.word = word;
        this.detail = detail;
    }
    public int getId() {
        return id;
    }
    public void setId(int id) {
        this.id = id;
    }
    public String getWord() {
        return word;
    }
    public void setWord(String word) {
        this.word = word;
    }
    public String getDetail() {
        return detail;
    }
    public void setDetail(String detail) {
        this.detail = detail;
    }
}
```

Android 单词显示的 Layout 设置 wordshow.xml。

```xml
<LinearLayout xmlns:android="http://schemas.android.com/apk/res/android"
    xmlns:tools="http://schemas.android.com/tools"
    android:layout_width="match_parent"
    android:layout_height="match_parent"
```

```xml
        android:background="@drawable/bg"
        android:orientation="vertical"
        android:textColor="@color/black_1" >
<TableLayout
        android:id="@+id/tablelayout01"
        android:layout_width="wrap_content"
        android:layout_height="wrap_content"
        android:collapseColumns="3"
        android:paddingLeft="35px"
        android:paddingRight="40px"
        android:paddingTop="40px"
        android:stretchColumns="1" >
<TextView
        android:layout_width="wrap_content"
        android:layout_height="wrap_content"
        android:text="@string/wordshow"
        android:textStyle="normal" />
<TableRow
        android:layout_width="wrap_content"
        android:layout_height="wrap_content" >
<TextView
        android:layout_width="wrap_content"
        android:layout_height="wrap_content"
        android:layout_centerHorizontal="true"
        android:layout_centerVertical="true"
        android:text="@string/word" />
<EditText
        android:id="@+id/word"
        android:layout_width="fill_parent"
        android:layout_height="wrap_content" />
</TableRow>
<TableRow
        android:layout_width="wrap_content"
        android:layout_height="wrap_content" >
<TextView
        android:layout_width="wrap_content"
        android:layout_height="wrap_content"
        android:layout_centerHorizontal="true"
```

```xml
        android:layout_centerVertical="true"
        android:text="@string/detail" />
<EditText
    android:id="@+id/detail"
    android:layout_width="fill_parent"
    android:layout_height="wrap_content" />
</TableRow>
</TableLayout>
<TableLayout
    android:layout_width="fill_parent"
    android:layout_height="wrap_content"
    android:gravity="center"
    android:orientation="horizontal" >
<TableRow
    android:layout_width="wrap_content"
    android:layout_height="wrap_content"
    android:gravity="center" >
<Button
    android:id="@+id/btnprior"
    android:layout_width="wrap_content"
    android:layout_height="wrap_content"
    android:text="@string/btnprior" />
<Button
    android:id="@+id/editbtn"
    android:layout_width="wrap_content"
    android:layout_height="wrap_content"
    android:layout_gravity="center_horizontal"
    android:text="@string/editWord" />
<Button
    android:id="@+id/btnnext"
    android:layout_width="wrap_content"
    android:layout_height="wrap_content"
    android:text="@string/btnnext" />
</TableRow>
</TableLayout>
<RatingBar
    android:id="@+id/ratingBar1"
    android:layout_width="wrap_content"
```

```xml
        android:layout_height="wrap_content"
        android:layout_gravity="center_horizontal" />
    <Button
        android:id="@+id/updatewordbtn"
        android:layout_width="match_parent"
        android:layout_height="wrap_content"
        android:text="@string/updateword" />
</LinearLayout>
```

Android 端单词显示界面的 WordShowActivity。WordShowActivity 中显示单词信息、单词下载、上一条单词、下一条单词查看，并将 JSON 数据插入到 SQLite 数据库并显示到界面当中。其中 SQLite 数据库已经在任务 2.5 介绍，具体内容参见任务 2.5 的 WordDBHelper。按钮事件处理如下。

```java
public class WordShowActivity extends Activity {
    WordDBHelper dbHelper;
    private EditText word,detail;
    private Button btnprior,btnnext,btnedit,updatewordbtn;
    Cursor c ;
    @Override
    protected void onCreate(Bundle savedInstanceState) {
        // TODO Auto-generated method stub
        super.onCreate(savedInstanceState);
        setContentView(R.layout.wordshow);
        dbHelper=new WordDBHelper(this);
        word=(EditText) findViewById(R.id.word);
        detail=(EditText) findViewById(R.id.detail);
        btnprior=(Button) findViewById(R.id.btnprior);
        btnnext=(Button) findViewById(R.id.btnnext);
        btnedit=(Button) findViewById(R.id.editbtn);
        updatewordbtn=(Button) findViewById(R.id.updatewordbtn);
        updatewordbtn.setOnClickListener(new OnClickListener() {
            @Override
            public void onClick(View arg0) {
                dbHelper.deleteWord();
                List<Word> words;
                try {
                    words = getLastJsonUser();
                    for(Word word:words){
                        dbHelper.insertDB(word.getWord(),
```

```
word.getDetail());
                    }
                    c = dbHelper.query();
                    if(c.getCount()>0){
                        c.moveToFirst();
                        word.setText(c.getString(1));
                        detail.setText(c.getString(2));
                    }
                    else
                        Toast.makeText(getApplicationContext(), "没有单词了!", 3000).show();
        Intent  intent=new  Intent(WordShowActivity.this,QueryActivity.class);
            startActivity(intent);
            } catch (Exception e) {
                    // TODO Auto-generated catch block
                    e.printStackTrace();
                }
            }
        });
        c = dbHelper.query();
        if(c.getCount()>0){
            c.moveToFirst();
            word.setText(c.getString(1));
            detail.setText(c.getString(2));
        }
        else
            Toast.makeText(getApplicationContext(), "没有单词了!", 3000).show();
        btnprior.setOnClickListener(new OnClickListener() {
            @Override
            public void onClick(View arg0) {
                if(c.moveToPrevious()){
                    word.setText(c.getString(1));
                    detail.setText(c.getString(2));}
                else
                {
                    Toast.makeText(getApplicationContext(), "已经是第
```

```
一条单词", 3000).show();
                    }
                }
            });
            btnnext.setOnClickListener(new OnClickListener() {
                @Override
                public void onClick(View v) {
                    // TODO Auto-generated method stub
                    if(c.moveToNext()){
                        word.setText(c.getString(1));
                        detail.setText(c.getString(2));}
                    else
                    {
                        Toast.makeText(getApplicationContext(), "已经是最后一条单词", 3000).show();
                    }
                }
            });
            btnedit.setOnClickListener(new OnClickListener() {
                @Override
                public void onClick(View v) {
                    Intent intent=new Intent(WordShowActivity.this, WordManage Activity.class);
                    Bundle b=new Bundle();
                    b.putString("word", word.getText().toString());
                    b.putString("detail", detail.getText().toString());
                    intent.putExtra("data",b);
                    startActivity(intent);
                }
            });
            dbHelper.close();
    }
```

数据显示参见 QueryActivity。通过 DBhelper 的 query()方法查询所有的单词,并利用 SimpleCursorAdapter 将单词信息显示到列表当中。

```
public class QueryActivity extends ListActivity {
    WordDBHelper dbhelper;
    @Override
    protected void onCreate(Bundle savedInstanceState) {
```

```java
            // TODO Auto-generated method stub
            super.onCreate(savedInstanceState);
            dbhelper=new WordDBHelper(this);
            Cursor c = dbhelper.query();
            String[] from = { "word", "detail"};
            int[] to = { R.id.textword, R.id.textdetail};
            SimpleCursorAdapter adapter = new SimpleCursorAdapter
(getApplicationContext(),
                    R.layout.row, c, from, to);
            ListView listView = getListView();
            listView.setAdapter(adapter);
            final AlertDialog.Builder builder = new AlertDialog.Builder(this);
            dbhelper.close();
        }
    }
```

PART 3 项目三 数独游戏

项目情境

数独游戏源自 18 世纪末瑞士的一种益智游戏，数独游戏盘面是个九宫，每一宫又分为 9 个小格。在这 81 格中给出一定的已知数字和解题条件，利用逻辑和推理，在其他的空格上填入 1~9 的数字。每个数字在每一行、每一列和每一宫中都只出现一次，且不重复。

本项目在手机上实现数独游戏设计，包括游戏首页设计、游戏介绍、九宫格界面设计、游戏中的数字选择、开始游戏和游戏过程控制、暂停游戏、继续游戏等功能，如图 3-1 所示。

图 3-1 数独游戏运行效果图

学习目标

- ☑ 巩固界面设计、事件处理。
- ☑ 掌握动态绘图，如画笔设置，绘制直线、矩形、椭圆、圆角矩形等。
- ☑ 掌握自定义对话框的设置。
- ☑ 进一步提高程序逻辑设计。
- ☑ 深入学习 Activity 的生命周期。
- ☑ 实现对音频文件的控制。

工作任务

任务名称
任务 3.1　数独游戏界面设计
任务 3.2　九宫格界面绘制
任务 3.3　数字键盘设计与实现
任务 3.4　游戏背景音乐设置
任务 3.5　继续游戏功能实现

任务 3.1　数独游戏界面设计

任务描述

本任务创建数独游戏的主界面，该界面用于启动和设置数独游戏，具体内容如下。

① 创建游戏首页界面，如图 3-2 所示，该页面包含的 4 个按钮分别是"开始玩游戏""关于游戏""继续玩游戏""退出游戏"，并实现每个按钮相应的功能。

② 实现"关于游戏"的游戏介绍功能。当单击"关于游戏"按钮时弹出 About 界面。

③ 实现"退出游戏"功能。当单击"退出游戏"按钮时，显示退出游戏的对话框。

④ 系统相关属性设置，设置两个菜单项内容为"设置属性"与"荣誉"，实现音乐背景与荣誉等级的设置。

图 3-2　数独游戏首页设置图

任务目标

① 通过任务巩固界面设计，如布局与常用控件的设置。
② 掌握 Activity 中应用对话框样式的设置方法。
③ 掌握对话框与菜单的定义。

任务分析

本任务主要实现游戏首页、关于游戏、退出游戏和游戏相关属性的设置功能。具体实现过程：

① 准备资源文件 strings.xml；
② 实现游戏首页的功能；
③ 实现"关于游戏"功能；
④ 实现"退出游戏"功能；
⑤ 实现"系统相关属性设置"功能；
⑥ 配置 AndroidManifest.xml 文件。

知识要点

- 常用组件的定义与加载。
- 事件的响应。
- menu 的定义与加载。
- 颜色资源与字符资源的使用。
- style 的定义与使用。
- Intent 的初步应用。
- AndroidManifest.xml 配置介绍。

任务实现

1. 创建字符串资源文件 strings.xml

在 res\values\路径中添加 strings.xml 文件，将任务中用到的字符串需在 strings.xml 文件中定义，便于工程中的其他文件引用该字符串。

```
<?xml version="1.0"encoding="utf-8"?>
<resources>
    <string name="hello">Hello World, SudokuActivity!</string>
    <string name="app_name">Sudoku</string>
    <string name="game_title">数独游戏</string>
    <string name="begin_text">开始玩游戏</string>
    <string name="continue_text">继续玩游戏</string>
```

```xml
<string name="about_text">关于游戏</string>
<string name="exit_text">退出游戏</string>
<string name="about_content">数独游戏简介,數獨游戏在 9×9 的方格内进行,分为 3x3 的小方格,每一横排每一竖排每一斜排都是 1~9 九个数字。每一横排每一竖排每一斜排不得重复任何一个数字。</string>
</resources>
```

2．实现游戏首页功能

游戏首页布局文件设计，创建 activity_main.xml，界面效果如图 3-2 所示。在 activity_main.xml 当中设计一个 TextView 和 4 个 Button。4 个按钮分别为开始玩游戏、关于游戏、继续玩游戏、退出游戏。

①activity_main.xml 代码如下。

```xml
<LinearLayout
xmlns:android="http://schemas.android.com/apk/res/android"
xmlns:tools="http://schemas.android.com/tools"
    android:layout_width="fill_parent"
    android:layout_height="fill_parent"
    android:background="#005566"
    android:orientation="vertical"
    android:padding="30dip">
<TextView
    android:id="@+id/mytext"
    android:layout_width="fill_parent"
    android:layout_height="wrap_content"
    android:gravity="center"
    android:text="@string/game_title "/>
//字符资源的引用
    <Button
    android:id="@+id/gamebeging"
    android:layout_width="fill_parent"
    android:layout_height="wrap_content"
    android:gravity="center"
    android:text="@string/ begin_text"
    android:textSize="25px"/>
<Button
    android:id="@+id/gamecontinue"
    android:layout_width="fill_parent"
    android:layout_height="wrap_content"
    android:gravity="center"
```

```xml
            android:text="@string/continue_text"
            android:textSize="25px"/>
    <Button
            android:id="@+id/gamesetting"
            android:layout_width="fill_parent"
            android:layout_height="wrap_content"
            android:gravity="center"
            android:text="@string/about_text "
            android:textSize="25px"/>
    <Button
            android:id="@+id/gameexit"
            android:layout_width="fill_parent"
            android:layout_height="wrap_content"
            android:gravity="center"
            android:text="@string/exit_text "
            android:textSize="25px"/>
</LinearLayout>
```

②创建 MainActivity，在该 Activity 中设置 activity_main.xml 为显示页面，依次加载 activity_main.xml 中的一个 TextView 和 4 个 Button。MainActivity 代码如下。

```java
public class MainActivity extends Activity {
    //定义菜单选项
    private staticfinalintITEM1 = Menu.FIRST;
    private staticfinalintITEM2 = Menu.FIRST+1;
    private TextView myText;
    //声明四个按钮
    private Button bt_begin,bt_continue,bt_exit,bt_about;
    @Override
    protected void onCreate(Bundle savedInstanceState) {
        super.onCreate(savedInstanceState);
        setContentView(R.layout.activity_main);
        //将开始游戏、继续游戏、关于、退出四个按钮实例化
        bt_begin=(Button)findViewById(R.id.gamebeging);
        bt_continue=(Button)findViewById(R.id.gamecontinue);
        bt_exit=(Button)findViewById(R.id.gameexit);
        bt_about=(Button)findViewById(R.id.gamesetting);
        //对四个按钮的事件实现监听
        bt_begin.setOnClickListener(this);
        bt_continue.setOnClickListener(this);
```

```
        bt_exit.setOnClickListener(this);
        bt_about.setOnClickListener(this);
        myText=(TextView)findViewById(R.id.mytext);
    }
    @Override
    public boolean onCreateOptionsMenu(Menu menu) {
        // Inflate the menu; this adds items to the action bar if it is present.
        getMenuInflater().inflate(R.menu.menu, menu);
        return true;
    }
}
```

3．实现"关于游戏"功能

"关于游戏"的界面设计。创建"关于游戏"的界面 about.xml，该界面中有一个文本框控件，用于介绍数独游戏的规则。

① about.xml 内容如下。

```
<?xmlversion="1.0"encoding="utf-8"?>
<LinearLayout
xmlns:android="http://schemas.android.com/apk/res/android"
    android:layout_width="fill_parent"
    android:layout_height="fill_parent"
    android:orientation="vertical">
<TextViewstyle="@style/style1"
    android:text="@string/about_content"
    android:layout_width="fill_parent"
    android:layout_height="fill_parent"
/>
</LinearLayout>
```

② 创建继承自 Activity 类的 About 类，该类将 about.xml 文件设置为显示界面。About 文件内容如下。

```
public class About extends Activity {
    @Override
    protected void onCreate(Bundle savedInstanceState) {
        super.onCreate(savedInstanceState);
        setContentView(R.layout.activity_about);
    }
}
```

③ 在 MainActivity 中通过按钮事件实现关于游戏的界面显示。首先将 MainActivity 实现

OnClickListener 接口,并实现事件单击方法 onClick(View arg0)。

MainActivity 实现接口的主要代码如下。

```
public class MainActivity extends Activity implements OnClickListener{…}
```

事件单击方法 onClick(View arg0)内容如下。

```
public void onClick(View arg0) {
    case R.id.gamesetting:
        Intent intent01=new Intent();
        intent01.setClass(this,About.class);
        startActivity(intent01);
        break;
}
```

4．实现"退出游戏"功能

实现"系统退出"功能,在游戏首页 MainActivity 中利用对话框实现退出。首先需要添加退出方法 exitGameDialog(),并在按钮单击事件处理方法 onClick(View arg0)添加事件响应的方法。

① 退出游戏方法 exitGameDialog()代码如下。

```
private void exitGameDialog() {
    // 弹出对话框
    AlertDialog.Builder exit_game=new AlertDialog.Builder(this);
    exit_game.setMessage("确定要退出").setCancelable(false);
    exit_game.setPositiveButton(" 确 定 ", new DialogInterface.
OnClickListener() {
        @Override
        public void onClick(DialogInterface dialog, int which) {
            // TODO Auto-generated method stub
            finish();
        }
    });
    exit_game.setNegativeButton(" 取 消 ", new DialogInterface.
OnClickListener() {            @Override
        public void onClick(DialogInterface dialog, int which) {
            // TODO Auto-generated method stub
            return;
        }
    });
    AlertDialog alert=exit_game.create();
    alert.show();
```

}

② 在按钮单击事件响应方法 onClick(View arg0)添加如下代码。

```
case R.id.gameexit:
        exitGameDialog();
        break;
```

5．实现"系统相关属性设置"功能

实现"系统相关属性设置"，创建菜单资源文件 menu.xml，设置两个选择项"设置属性"与"荣誉"，并在 MainActivity 加载 Menu 菜单，响应菜单事件。

① 在 res\menu\路径中创建选项菜单 menu.xml，并在 MainActivity 中加载两个菜单项。

```xml
<?xml version="1.0"encoding="utf-8"?>
<menu xmlns:android="http://schemas.android.com/apk/res/android">
<item
    android:id="@+id/setting"
    android:title="设置属性"
    android:icon="@drawable/ic_launcher"/>
<item
    android:id="@+id/readRecord"
    android:title="荣誉"/>
</menu>
```

② 在创建 MainActivity 时会自动创建 onCreateOptionsMenu(Menu menu) 方法，该方法的作用是将 R.menu.menu 设置为本 MainActivityActivity 的选项菜单。

```java
/*从 xml 定义的菜单资源中生成一个菜单*/
    @Override
    public boolean onCreateOptionsMenu(Menu menu) {
        // TODO Auto-generated method stub
        super.onCreateOptionsMenu(menu);
        MenuInflater inflater=getMenuInflater();
        inflater.inflate(R.menu.menu, menu);
        return true;
    }
```

③ 实现菜单单击事件。在 onOptionsItemSelected(MenuItem item) 方法中实现菜单单击事件，方法中利用 switch 实现菜单事件响应。

```java
@Override
public boolean onOptionsItemSelected(MenuItem item) {
    switch(item.getItemId()){
    case R.id.setting:
        Intent intent=new Intent(this,Prefs.class);
```

```
            startActivity(intent);
            break;
        case R.id.readRecord:
            break;
        default:
            return super.onOptionsItemSelected(item);
    }
    return false;
}
```

6．配置 AndroidManifest.xml 文件

在 AndroidManifest 文件中实现 MainActivity 与 About 两个 Activity 文件的注册。其中 MainActivity 设置为启动 Acitivity。此外，About 需用自定义样式实现。详细内容如下所示。

① 在 res\values\styles.xml 中添加如下代码定义自定义样式，表示该 Activity 的样式为对话框样式。styles.xml 文件内容如下。

```
<style name="style1">
    <item name="android:textSize">20sp</item>
    <item name="android:text">#EC9237</item>
</style>
    <style name="style_window"parent="android:style/Theme.Dialog">
    <item name="android:textSize">20sp</item>
    <item name="android:text">#EC9237</item>
</style>
```

② AndroidManifest.xml 文件内容如下。

```
<?xml version="1.0"encoding="utf-8"?>
<manifest xmlns:android="http://schemas.android.com/apk/res/android"
    package=" com.example.sudoku "
    android:versionCode="1"
    android:versionName="1.0">
<uses-sdk
    android:minSdkVersion="8"
    android:targetSdkVersion="17"/>
<application
    android:allowBackup="true"
    android:icon="@drawable/ic_launcher"
    android:label="@string/app_name"
>
<activity
```

```
        android:name=" com.example.sudoku.MainActivity"
        android:label="@string/app_name">
        <intent-filter>
            <actionandroid:name="android.intent.action.MAIN"/>
            <categoryandroid:name="android.intent.category.LAUNCHER"/>
        </intent-filter>
    </activity>
    <activity
        android:name="com.example.sudoku.About"
        android:label="@string/title_activity_about"
        android:theme="@style/style_window">
    </activity>
</application>
</manifest>
```

任务 3.2　九宫格界面绘制

任务描述

绘制九宫格。利用 9 条横线和 9 条竖线，把屏幕分成 9×9 的 81 个格子，并对每个格子添加触屏事件响应，当手指点击屏幕上某个格子，选中一个矩形，填入初始的数字，其页面效果如图 3-3 所示。

任务目标

① 理解自定义视图的设计思想，掌握自定义视图 View 的定义方法。

② 掌握 2D 图形的绘制，如画笔设置、字体设置、颜色设置等，并能绘制线条、矩形、圆形等形状。

③ 掌握触屏事件的处理。

图 3-3　数独游戏九宫格界面

任务分析

本任务主要通过自定义视图实现界面设计,并在 onDraw 方法中实现九宫格的绘制。具体实现过程:

① 绘制九宫格界面;
② 绘制选中的矩形框;
③ 结合任务 3.1 实现九宫格界面显示。

知识要点

1. 动态图形绘制

基本思路:创建一个类继承 View 类或者继承 SurfaceView 类覆盖 onDraw 方法,使用 Canvas 对象在界面上面绘制不同的图形,使用 invalidate()方法刷新界面。

画布——Canvas
画笔——Paint
颜色——Color
画线的路径——Path

2. Canvas 类

使用该类提供的各种方法可以在画布上绘制线条、矩形、圆形等图形。Android 中,屏幕是由 Activity 类的对象支配的,Activity 类的对象引用 View 类(布局)对象,View 类的对象又引用 Canvas 类的对象,用户通过重写 View 类的 onDraw 方法,在指定的画布上绘图。

3. Paint 类

Paint 代表画笔类,包含图形的样式、颜色、线条粗细以及绘制图形所需要的其他信息,使用该类的方法,可设置所画图形的各种属性,如颜色、样式等。

```
Paint cPaint = new Paint();
cPaint.setColor(Color.BLUE);
```

4. Color 类

Android 中使用 4 个数字来表示颜色,分别是透明度(Alpha)、红(Red)、绿(Green)、蓝(Blue)四个颜色值(ARGB)。每个数字取值 0~255,因此一个颜色可以用一个整数来表示。为了运行效率,Android 编码时用整数 Color 类实例来表示颜色。

红、绿、蓝三个值是就是代表颜色的取值,Alpha 最低值为 0,表示颜色完全透明,最高可取值为 255,表示颜色完全不透明。如果需要颜色透明、半透明,可以取值 0~255 的一些值。

有下面几种方式来创建或表示一个颜色:

(1)使用 Color 类的常量。

```
Int color = Color.BULE;    // 创建一个蓝色
```

（2）如果知道ARGB的取值，可以使用Color类的静态方法argb创建一个颜色。

```
Int color = Color.argb(127,255,0,255);  // 半透明的紫色
```

（3）使用XML资源文件表示颜色是一个扩展性比较好的方式，便于修改颜色值。

```xml
<?xml version="1.0"encoding="utf-8">
<resources>
    <color name="mycolor">#7fff00ff</color>
</resources>
```

定义一个名为mycolor颜色，可以引用mycolor来获取该颜色值。Java代码中可以使用ResourceManager类中的getColor来获取该颜色。Java样例代码如下：

```
Int color = getResources().getColor(R.color.mycolor);
```

这与第二种方法得到的值一样。getResources()方法返回当前活动Activity的ResourceManager类实例。

5．Path类

定义一组矢量绘图路径，表示要画的图形的轨迹和位置，该类包含一组矢量绘图命令，可以绘制线条、矩形、圆形等图形。

任务实现

1．绘制九宫格界面

创建puzzleView，绘制九宫格界面，首先通过onSizeChanged方法获取整体界面的宽和高，然后用整体界面的宽和高除以9来计算每个格子的宽和高，并赋值给width和height。

在onDraw方法中定义线条颜色，通过for循环绘制线条，线条绘制方法drawLine()。

（1）绘制自定义视图

创建puzzleView类，该类需继承于View类。在puzzleView类中重写onDraw()方法。

（2）绘制线条与数据填充

在onDraw方法中创建Paint画笔对象，实现绘制九宫格的功能。绘制9×9的格子，利用线条颜色的不同，把81个格子分成9个九宫格。

- 创建puzzleView，绘制九宫格界面，首先通过onSizeChanged方法获取整体界面的宽和高，然后用整体界面的宽和高除以9来计算每个格子的宽和高，并赋值给width和height。
- 在onDraw方法中定义线条颜色通过for循环绘制线条，线条绘制方法drawLine()。
- 创建二维数组，将初始数据填入相应的格子中定义二维数组，通过双重for循环实现二维数组的绘制。

```
import android.content.Context;
import android.graphics.Canvas;
import android.graphics.Color;
import android.graphics.Paint;
import android.graphics.Rect;
```

```java
import android.util.AttributeSet;
import android.view.MotionEvent;
import android.view.View;

public class PuzzleView extends View {
    private Game new_game;
    private int difficulty;
    private int i,j,m=0;
    private float width,height;
    private int selX,selY;
    public String[][]  puzzle = new String[][]{
            {"1"," "," "," "," "," "," "," "," "},
            {"2"," "," "," "," "," "," "," "," "},
            {"3"," "," "," "," "," "," "," "," "},
            {"4"," "," "," "," "," "," "," "," "},
            {"5"," "," "," "," "," "," "," "," "},
            {"6"," "," "," "," "," "," "," "," "},
            {"7"," "," "," "," "," "," "," "," "},
            {"8"," "," "," "," "," "," "," "," "},
            {"9"," "," "," "," "," "," "," "," "},
    };
    public PuzzleView(Context context) {
        // TODO Auto-generated constructor stub
        super(context);
        //取得绘图的上下文环境
        new_game=(Game)context;
        setFocusable(true);//允许键盘事件为true，才会响应该事件
        setFocusableInTouchMode(true);//允许触屏事件为true才会响应相应事件
    }
    /*
     * onSizeChanged 在屏幕发生改变的时候调用，在初始化一个屏幕时，在onCreate方法
之前调用通过重写该方法，在其内部获取屏幕的宽度，从而获取游戏中矩形框的宽度和高度
     * */
    @Override
    protected void onSizeChanged(int w, int h, int oldw, int oldh) {
        // TODO Auto-generated method stub
        super.onSizeChanged(w, h, oldw, oldh);
        width=w/9f;
```

```
            height=h/9f;
    }
    //绘制9*9的格子，利用线条颜色的不同把81个格子形成九宫格
    @Override
    protected void onDraw(Canvas canvas) {
        // TODO Auto-generated method stub
        super.onDraw(canvas);
        /*绘制背景颜色*/
        Paint backgroundpaint=new Paint();
        backgroundpaint.setColor(getResources().getColor(R.color.game_background));
        canvas.drawRect(0, 0, getWidth(), getHeight(), backgroundpaint);
        //定义颜色
        Paint dark = new Paint();
        dark.setColor(getResources().getColor(R.color.puzzle_dark));
        Paint light = new Paint();
        light.setColor(getResources().getColor(R.color.puzzle_light));
        Paint hilight = new Paint();
        hilight.setColor(getResources().getColor(R.color.puzzle_hilite));
        Paint yewllow = new Paint();
        yewllow.setColor(Color.YELLOW);
        //绘制线条
        for(i=0;i<=9;i++){
            canvas.drawLine(0,i*height,getWidth(),i*height,light);
            canvas.drawLine(0,i*height+1,getWidth(),i*height+1,hilight);
            canvas.drawLine(i*width,0,i*width,getHeight(),light);
            canvas.drawLine(i*width+1,0,i*width+1,getHeight(),hilight);
        }
        for(i=0;i<=9;i=i+3){
            canvas.drawLine(0,i*height,getWidth(),i*height,yewllow);
            canvas.drawLine(0,i*height+1,getWidth(),i*height+1,hilight);
            canvas.drawLine(i*width,0,i*width,getHeight(),yewllow);
            canvas.drawLine(i*width+1,0,i*width+1,getHeight(),hilight);
        }
        //将初始数据填入相应的格子中定义二维数组，通过双重for循环实现二维数组的绘制。绘制九宫格内数字
        Paint frontPaint=new Paint();
        frontPaint.setColor(Color.WHITE);
```

```
            frontPaint.setTextSize(25);
            for(i=0;i<9;i++)
                for(j=1;j<9;j++){
                    canvas.drawText(puzzle[i][j],   j*width+(width*0.3f),
i*height+ (height*0.6f),frontPaint);
                }
    }
}
```

2．实现矩形绘制

矩形的绘制需要首先定义一个矩形对象，在 onSizeChanged 更新矩形大小，根据触屏事件即时更新矩形的位置坐标。

① 在 PuzzleView 类中声明矩形对象。

```
private final Rect selRect=new Rect();
```

② 添加 getRect(float selX,float selY,Rect r)方法，该方法根据参数中的 selX 与 selY，更新矩形 r 的位置。

```
/*根据传入的参数 selX 和 selY 更新矩形位置*/
private void getRect(float selX,float selY,Rect r){
    r.set((int)(selX*width+1),(int)(selY*height+1),(int)(selX*width+
width-1),(int)(selY*height+height-1));
}
```

③ 在 onSizeChanged 方法调用 getRect(selX,selY,selRect)方法，这样可以在屏幕发生旋转时及时更新矩形的位置。onSizeChanged 在屏幕发生改变的时候调用，在初始化一个屏幕时，在 onCreate 方法之前调用通过重写该方法，在其内部获取屏幕的宽度，从而获取游戏中矩形框的宽度和高度。

```
@Override
protected void onSizeChanged(int w, int h, int oldw, int oldh) {
    // TODO Auto-generated method stub
    super.onSizeChanged(w, h, oldw, oldh);
    width=w/9f;
    height=h/9f;
    getRect(selX, selY, selRect);
}
```

④ 在 onDraw 方法中绘制矩形。

```
Paint selPaint=new Paint();
selPaint.setColor(R.color.puzzle_selected);
canvas.drawRect(selRect, selPaint);
```

⑤ 添加 selectXY(int x,int y)方法，根据用户点击的方向键，设置选中矩形的位置，

invalidate( )通知 onDraw 重新绘图。

```
        public void selectXY(int x,int y){
            invalidate(selRect);
            if(y==-1) {selY=8;}
            else if(y==9) {selY=0;}
            else selY=y;
            if(x==-1) {selX=8;
            }else if(x==9){selX=0;
            }else selX=x;
            //更新矩形的长、宽、左上角坐标等属性
            getRect(selX,selY,selRect);
            //重新绘制矩形
            invalidate(selRect);
        }
```

⑥ 添加触屏事件处理代码,获取触屏事件的位置,并调用 selXY()方法更新矩形框的位置。

```
/*相应触屏功能*/
@Override
public boolean onTouchEvent(MotionEvent event) {
    // TODO Auto-generated method stub
    //检测事件类型
    if(event.getAction()!=MotionEvent.ACTION_DOWN){
        return super.onTouchEvent(event);
    }
    //计算触屏的位置在哪个矩形上
    int x=(int)(event.getX()/width);
    int y=(int)(event.getY()/height);
    selectXY(x,y);
    return false;
}
```

3. 结合任务 3.1 实现九宫格界面显示

结合任务 3.1 实现九宫格界面显示,在任务 3.1 中的游戏首页中单击"开始游戏"显示九宫格界面。

首先创建继承自 Acitvity 的 Game 类,调用 setContentView()方法将 Game 的显示界面设置为 puzzleView,并在 MainActivity 的 onClick 事件中添加 Intent 实现页面跳转。

① MainActivity 中的事件监听代码如下。

```
beginGame.setOnClickListener(this);
//事件处理
```

```
case R.id.gamebeging:
        startNewGame();
        break;
......
private void startNewGame(){
        Intent intent=new Intent(this,Game.class);
        startActivity(intent);
    }
......
```

② Game 代码定义如下。

```
package com.example.sudoku;
import android.app.Activity;
import android.os.Bundle;

public class Game extends Activity {
    protected PuzzleView puzzleview;
    public String continueString;
    @Override
    protected void onCreate(Bundle savedInstanceState) {
        // TODO Auto-generated method stub
        super.onCreate(savedInstanceState);
        puzzleview=new PuzzleView(this);
        //setContentView就是设置一个Activity的显示界面,设置该Activity的显示
界面为puzzleview
        setContentView(puzzleview);
        //请求puzzleview获得焦点
        puzzleview.requestFocus();
    }
}
```

③ 将 Game 在 AndroidManifest.xml 中注册，代码如下。

```
<activity
        android:name="com.example.soduku.Game"
        android:label="@string/app_name"
        >
```

技能训练

任务 1：绘制线条，画出图 3-4 所示的网格，9×9 单元格。

任务 2：利用 2D 图形绘制画出图 3-4 所示的奥运五环，并且附上相应的文字。

任务 3：利用动态图形绘制图 3-4 的第三个图，结合触屏事件实现涂鸦功能。

图 3-4　程序运行效果图

任务 3.3　数字键盘设计与实现

任务描述

本任务是实现小键盘的数字输入并进行逻辑判断，基本功能为单击游戏界面矩形，弹出小键盘界面，游戏玩家选择数字，系统对数字进行逻辑判断，如图 3-5 所示。

数字输入判断规则为：不能在同行、同列、同一个九宫格之内输入相同的数字。

任务目标

① 掌握自定义对话框的设计与实现。

② 掌握 Context 上下文传递。

③ 掌握数组内的数据比较方法。

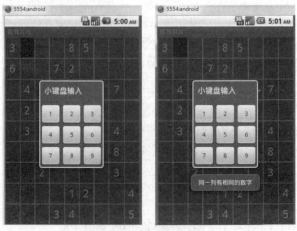

图 3-5　数字键盘效果图

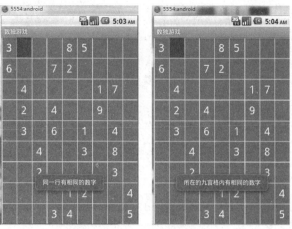

图 3-5 数字键盘效果图（续）

任务分析

本任务主要实现数字的输入和对输入数字的验证，数字验证规则是"不能在同行、同列、同一个九宫格之内输入相同的数字"。符合规则的数字允许写入对应的数组当中，否则提示错误。具体实现过程：

① 数字键盘设计与实现；
② 实现输入的数字验证的方法；
③ 在键盘触屏事件中实现数字验证。

知识要点

① 自定义对话框设计。
② 对输入数字的逻辑判断方法。
③ KeyPad 类中键盘响应事件。
④ KeyPad 类中的按钮单击事件。

任务实现

1．数字键盘设计与实现

数字键盘设计与实现需要实现三步，分别是数字键盘页面设计、数字键盘对话框类的设计、从 MainActivity 中显示数字键盘。下面依次介绍其实现过程。

① 数字键盘的页面设计。创建 keypad.xml 实现数字键盘的界面设计，该界面中有三行三列按钮控件，用于放置 1~9 的 9 个数字。

```
<?xml version="1.0" encoding="utf-8"?>
<TableLayout xmlns:android="http://schemas.android.com/apk/res/android"
android:layout_width="wrap_content "
android:layout_height=wrap_content">
```

```xml
<TableRow>
    <Button android:id="@+id/keypad_1"android:text="1"android:layout_width="50dp"/>
    <Button android:id="@+id/keypad_2"android:text="2"android:layout_width="50dp"/>
    <Button android:id="@+id/keypad_3"android:text="3"android:layout_width="50dp"/>
</TableRow>
<TableRow>
    <Button android:id="@+id/keypad_4"android:text="4"android:layout_width="50dp"/>
    <Button android:id="@+id/keypad_5"android:text="5"android:layout_width="50dp"/>
    <Button android:id="@+id/keypad_6"android:text="6"android:layout_width="50dp"/>
</TableRow>
<TableRow>
    <Button android:id="@+id/keypad_7"android:text="7"android:layout_width="50dp"/>
    <Button android:id="@+id/keypad_8"android:text="8"android:layout_width="50dp"/>
    <Button android:id="@+id/keypad_9"android:text="9"android:layout_width="50dp"/>
</TableRow>
</TableLayout>
```

② 创建一个自定义对话框类 KeyPad，该类继承自 Dialog 类。设置其显示界面为 keypad.xml，在该类的构造方法中获取上下文环境，并将其父视图设置为 PuzzleView 的对象，使用 findViewById ( )方法依次加载 keypad.xml 当中的 9 个按钮。

```java
import android.app.Dialog;
import android.content.Context;
import android.os.Bundle;
import android.view.View;
import android.widget.Button;
public class Keypad extends Dialog implements android.view.View.OnClickListener {
    private PuzzleView fatherView;
    private Button[] key=new Button[9];
    public Keypad(Context context, PuzzleView father) {
        super(context);
```

```
            // TODO Auto-generated constructor stub
        fatherView=father;
    }
    privatevoid findButtonId(){
        key[0]=(Button)findViewById(R.id.keypad_1);
        key[1]=(Button)findViewById(R.id.keypad_2);
        key[2]=(Button)findViewById(R.id.keypad_3);
        key[3]=(Button)findViewById(R.id.keypad_4);
        key[4]=(Button)findViewById(R.id.keypad_5);
        key[5]=(Button)findViewById(R.id.keypad_6);
        key[6]=(Button)findViewById(R.id.keypad_7);
        key[7]=(Button)findViewById(R.id.keypad_8);
        key[8]=(Button)findViewById(R.id.keypad_9);
    }
}
```

③ 在触屏单击事件实现小键盘显示，在 PuzzleView 类中的 onTouchEvent( )方法中通过 new KeyPad(getContext(), this).show()方法实现数字键盘显示。

```
new KeyPad(getContext(), this).show();
```

2．实现验证数字输入的方法

在 PuzzleView 当中添加 setSelectTitle(String d)方法，setSelectTitle(String d)方法用于实现对输入数字的逻辑判断。对用户的输入进行合法性检查，提示错误信息或者将合法的数据写入 puzzle 数组，并通知 onDraw 方法重新绘图，setSelectTitle(String d)方法的实现逻辑思路。

① 检查同一列是否有相同的数字。

② 检查同一行是否有相同数字。

③ 检查所在的九宫格是否有相同数字。

④ 判断输入的数字是否为 0，如果为 0 则将其重置为空；如果 info 不为空则显示 info 内容，为空则将该矩形框的 puzzle 数组的内容重置为 d，并重新绘制图形。

⑤ 验证 puzzle 数组内是否存在空值，如果存在，将 finishflag 赋值为 1。

⑥ 根据 finishflag 的值判断游戏是否结束，如果为 0 则通过，否则继续游戏。

```
public void setSelectTitle(String d)
{
    int row,col;
    int finishflag = 0;
    String info = "";
    /*1. 检查同一列是否有相同的数字 */
    for(row = 0;row<9;row++)
    {
```

```
            if(puzzle[selX][row].equals(d))
            {
                info = "同一列有相同的数字";
    break;
            }
        }
        /*2.检查同一行是否有相同数字*/
        for(row=0;row<9;row++)
        {
            if(puzzle[row][selY].equals(d))
            {
                info = "同一行有相同的数字";
                break;
            }
        }
        /*3.检查所在的九宫格是否有相同数字*/
        for(row = (selX/3)*3;row<=(selX/3*3+2);row++)
            for(col = (selY/3)*3;col<=(selY/3)*3+2;col++)
            {
                if(puzzle[row][col].equals(d))
                {
                    info = "所在的九宫格内有相同的数字";
                    break;
                }
            }
```

//4.判断输入的数字是否为 0，如果为 0 则将其重置为空，并将该矩形框的 puzzle 数组的内容重置为 d，重新绘制图形，如果 info 不为空则显示 info 内容。

```
        if(info.equals(""))
        {
            if(d.equals("0"))
                d ="";
            puzzle[selX][selY] = d;
            invalidate(selRect);
        }
        else
        {
        Toast.makeText(this.new_game.getApplicationContext(), info, Toast.LENGTH_SHORT).show();
```

```
        }
    //5.验证puzzle数组内是否存在空值,如果存在将finishflag赋值为1
        for(row=0;row<9;row++)
            for(col=0;col<9;col++)
            {
                if(puzzle[row][col].equals(" "))
                {
                    finishflag = 1;
                }
            }
    //6.根据finishflag的值判断游戏是否结束,如果为0则通过,否则继续游戏
        if(finishflag == 0)
        {
    Toast.makeText(this.new_game.getApplicationContext(), "恭喜你通关啦",
Toast.LENGTH_SHORT).show();
        }
    }
```

3．在键盘触屏事件中实现数字验证

① 在 **PuzzleView** 类当中添加键盘响应事件,在每个按钮事件中调用 setSelectTitle(String d) 方法实现键盘输入数字判断,并通过 selectXY()实现矩形框的选择。

```
@Override
    public boolean onKeyDown(int keyCode, KeyEvent event) {
        // TODO Auto-generated method stub
        switch(keyCode){
        case KeyEvent.KEYCODE_DPAD_UP:
            selectXY(selX,selY-1);
            break;
        case KeyEvent.KEYCODE_DPAD_DOWN:
            selectXY(selX,selY+1);
            break;
        case KeyEvent.KEYCODE_DPAD_LEFT:
            selectXY(selX-1,selY);
            break;
        case KeyEvent.KEYCODE_DPAD_RIGHT:
            selectXY(selX+1,selY);
            break;
        case KeyEvent.KEYCODE_0:
```

```
            setSelectTitle ("0");
            break;
        case KeyEvent.KEYCODE_1:
            setSelectTitle("1");
            break;
        case KeyEvent.KEYCODE_2:
            setSelectTitle ("2");
            break;
        case KeyEvent.KEYCODE_3:
            setSelectTitle ("3");
            break;
        case KeyEvent.KEYCODE_4:
            setSelectTitle ("4");
            break;
        case KeyEvent.KEYCODE_5:
            setSelectTitle ("5");
            break;
        case KeyEvent.KEYCODE_6:
            setSelectTitle ("6");
            break;
        case KeyEvent.KEYCODE_7:
            setSelectTitle ("7");
            break;
        case KeyEvent.KEYCODE_8:
            setSelectTitle ("8");
            break;
        case KeyEvent.KEYCODE_9:
            setSelectTitle ("9");
            break;
        default:
            return super.onKeyDown(keyCode, event);
        }
        return true;
    }
```

② 在 KeyPad 界面中实现数字按钮单击事件处理，在 onCreate(Bundle savedInstanceState) 方法在依次调用 fathview 当中的 setSelectTitle (String d)方法对输入的数据进行逻辑判断。

```java
@Override
protected void onCreate(Bundle savedInstanceState) {
    // TODO Auto-generated method stub
    super.onCreate(savedInstanceState);
    setTitle("小键盘输入");
    setContentView(R.layout.keypad);
    findButtonId();
    for(int i=0;i<9;i++){
        key[i].setOnClickListener(this);
    }
}
@Override
public void onClick(View v) {
    // TODO Auto-generated method stub
    switch(v.getId())
    {
    case R.id.keypad_1:
        fatherView. setSelectTitle("1");
    case R.id.keypad_2:
        fatherView. setSelectTitle ("2");
        break;
    case R.id.keypad_3:
        fatherView. setSelectTitle ("3");
        break;
    case R.id.keypad_4:
        fatherView. setSelectTitle ("4");
        break;
    case R.id.keypad_5:
        fatherView. setSelectTitle ("5");
        break;
    case R.id.keypad_6:
        fatherView. setSelectTitle ("6");
        break;
    case R.id.keypad_7:
        fatherView. setSelectTitle ("7");
        break;
    case R.id.keypad_8:
        fatherView. setSelectTitle ("8");
```

```
            break;
        case R.id.keypad_9:
            fatherView. setSelectTitle ("9");
            break;
        default:
            fatherView. setSelectTitle (" ");
            break;
        }
        dismiss();
    }
}
```

任务 3.4 游戏背景音乐设置

任务描述

结合 PreferenceScreen 对游戏参数进行设置，并在游戏过程中设置游戏的背景音乐和游戏音乐，效果如图 3-6 所示。

图 3-6 背景音乐设置图

任务目标

① 掌握 PreferenceScreen 与 PreferenceActivity 的定义设置。
② 掌握音频文件的控制方法。

任务分析

本任务主要实现背景音乐的选项设置，背景音乐及游戏过程音效的控制。具体实现过程：
① 准备音频文件；
② 游戏参数设置；

③ 游戏背景音乐和游戏音效的控制；

④ 配置 AndroidManifest.xml 文件。

知识要点

1. PreferenceScreen 的应用

PreferenceScreen 用于偏好设置，比如手机的"设置"面板。PreferenceScreen 代表一个偏好设置的根节点，在 PreferenceScreen 中可以定义各种选项控件，如单选按钮、复选按钮、列表等选择控件。

PreferenceScreen 设置的基本步骤如下。

① 在项目资源文件中新建 prefs.xml 文件，路径为 res\xm\prefs.xml。

② 根元素为 PreferenceScreen 代表显示整个屏幕。内部嵌套 PreferenceCategory 标签，表示偏好类别，在 PreferenceCategory 标签内部可以根据需要设置复选框、输入框、列表等显示控件。

③ 利用 Java 代码实现 PreferenceScreen 的设置。将 PreferenceScreen 加载到 Acitivity 中有两种方法。

- 方法一：创建类继承自 PreferenceActivity，在 onCreate 中添加 addPreferencesFromResource(R.xml. prefs)方法。注意该方法适用于 android2.3。
- 方法二：创建类继承自 PreferenceFragment，在 onCreate 中添加 addPreferencesFromResource(R.xml. prefs)方法。通过 FragmentManager 对 PreferenceFragment 对象实现管理和控制。该方法将在任务拓展中详细展开。

2. 音乐文件的设置

Android 中提供了音频和视频的控制类 MediaPlayer，该类提供了播放、暂停、停止和重复播放等方法。该类位于 android.media 包下。除了 MediaPlayer 类之外 SoundPool 也是对声频文件进行控制的类，下面对这两个类进行分析介绍。

① MediaPlayer：此类适合播放较大文件，应该存储在 SD 卡上，而不是在资源文件里，且每次只能播放一个音频文件。可以通过 4 种方式播放多媒体文件，分别是从资源文件中播放、从文件系统播放、通过 URI 的方式从网络播放、通过设置数据源的方式，相关代码如下。

- 从资源文件中播放。

```
MediaPlayerplayer =  new MediaPlayer.create(this,R.raw.test);
player.stare();
```

- 从文件系统播放。

```
MediaPlayer  player  =   new MediaPlayer();
player.setDataSource("/sdcard/test.mp3");
player.prepare();
player.start();
```

- 通过 URI 的方式从网络播放。

```
String path="http://*************.mp3";      //歌曲的网络地址
Uri  uri = Uri.parse(path);
MediaPlayer  player = new MediaPlayer.create(this,uri);
player.start();
```

- 通过设置数据源的方式。

```
MediaPlayer  player = new MediaPlayer.create();
String path="http://*************.mp3";      //歌曲的网络地址
player.setDataSource(path);
player.prepare();
player.start();
```

② SoundPool 类的特点就是低延迟播放，适合播放实时音频实现同时播放多个声音，如游戏中炸弹的爆炸音等小资源文件。此类音频比较适合放到 res/raw 资源文件夹下和程序一起打包生成 APK 文件。其用法如下。

```
SoundPool soundPool = new SoundPool(4, AudioManager.STREAM_MUSIC, 100);
HashMap<Integer, Integer> soundPoolMap = new HashMap<Integer, Integer>();
soundPoolMap.put(1, soundPool.load(this, R.raw.dingdong1, 1));
soundPoolMap.put(2, soundPool.load(this, R.raw.dingdong2, 2));
public void playSound(int sound, int loop) {
  AudioManager mgr = (AudioManager)this.getSystemService(Context.AUDIO_SERVICE);
    float streamVolumeCurrent = mgr.getStreamVolume(AudioManager.STREAM_MUSIC);
    float streamVolumeMax = mgr.getStreamMaxVolume(AudioManager.STREAM_MUSIC);
    float volume = streamVolumeCurrent/streamVolumeMax;
    soundPool.play(soundPoolMap.get(sound), volume, volume, 1, loop, 1f);
    //参数:1.Map 中取值   2.当前音量   3.最大音量  4.优先级   5.重播次数   6.播放速度
  }
  this.playSound(1, 0);
```

任务实现

1. 为游戏准备音频文件

在 res 中添加 raw 文件夹并在其中添加音乐文件 music.mp3 和 yue.mp3，位置如图 3-7 所示。

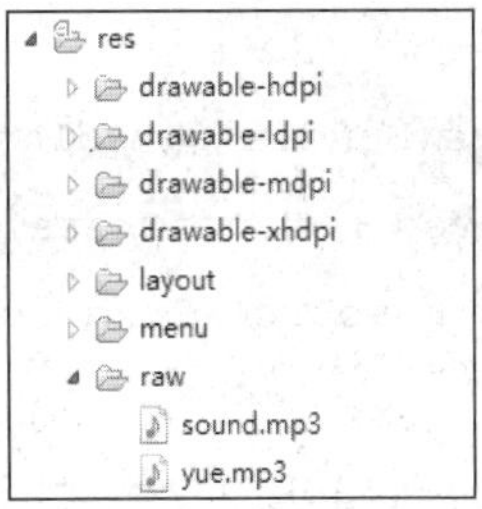

图 3-7　音乐文件

2．游戏参数设置

① 在 res 中添加 prefs.xml 文件，用于控制音频开关状态。prefs.xml 文件内容如下。

```xml
<?xml version="1.0" encoding="utf-8"?>
<PreferenceScreen
xmlns:android="http://schemas.android.com/apk/res/android">
    <PreferenceCategory android:title="基本设置">
    <CheckBoxPreference
    android:key="music"
    android:summaryOn="播放"
    android:summaryOff="关闭"
    android:defaultValue="true"
    android:title="背景音乐设置"
    />
    <CheckBoxPreference
    android:key="sound"
    android:summaryOn="音效开"
    android:summaryOff="音效关"
    android:defaultValue="true"
    android:title="游戏音乐设置"
    />
</PreferenceCategory>
</PreferenceScreen>
```

② 在 com.example.sudoku 中定义继承自 PreferenceActivity 的类 Prefs.java，在该类实现音频文件开关状态控制。

```java
package com.example.sudoku;
import android.R.bool;
import android.content.Context;
import android.os.Bundle;
import android.preference.PreferenceActivity;
import android.preference.PreferenceManager;
public class Prefs extends PreferenceActivity{
```

```
    @Override
    protected void onCreate(Bundle savedInstanceState) {
        super.onCreate(savedInstanceState);
        addPreferencesFromResource(R.xml.prefs);
    }
    public static boolean getBackMusic(Context context){
        return PreferenceManager.getDefaultSharedPreferences(context).getBoolean("music",true);
    }
    public static boolean getSoundSet(Context context){
        return PreferenceManager.getDefaultSharedPreferences(context).getBoolean("sound",true);
    }
}
```

③ 在 MainActivity 的菜单选择事件 onOptionsItemSelected(MenuItem item)中通过 Intent 实现 Prefs 显示，相关代码如下。

```
@Override
public boolean onOptionsItemSelected(MenuItem item) {
    switch(item.getItemId()){
    case R.id.setting:
        Intent intent=new Intent(this,Prefs.class);
        startActivity(intent);
        break;
    case R.id.readRecord:
        break;
    default:
        return super.onOptionsItemSelected(item);
    }
    return false;
}
```

3．游戏背景音乐和游戏音效的控制

在 com.example.sudoku 中创建 music.java 用于定义音频文件的加载、播放与暂停。并在 music 类添加 paly()和 stop()方法用于控制音频文件的加载播放暂停。

① 结合 Game 中 Activity 的暂停（onPause）、唤醒（onResume）、停止（onStop）调用 music 的 play()和 stop()方法实现音频的播放与暂停。

```
package com.example.sudoku;
```

```java
import android.content.Context;
import android.media.MediaPlayer;
public class music {
    private static MediaPlayer mp=null;
    private static MediaPlayer mmsound=null;
    /*
     * 控制启动音乐播放,context 是音乐播放时的上下文环境,
     * resources 是要播放的音乐资源文件
     * */
    public static void paly(Context context,int resources){
        stop(context);//播放指定音乐之前先把上下文任意多媒体停止
        if(Prefs.getBackMusic(context)){
            mp=MediaPlayer.create(context, resources);
            mp.setLooping(true);
            mp.start();
        }
    }
    public static void stop(Context context) {
        if(mp!=null){
            mp.stop();
            mp.release();
            mp=null;
        }
    }
    public static void palySound(Context context,int resource){
        stop(context);
        if(Prefs.getSoundSet(context)){
            mmsound=MediaPlayer.create(context, resource);
            mmsound.start();
        }
    }
}
```

② 在 Game 中结合 Activity 的启动、暂停方法,实现音频的控制。当暂停游戏时,音频停止播放,当重新启动游戏时,音频文件开始播放,具体代码在 Game 中修改如下。

```java
@Override
protected void onResume() {
    // TODO Auto-generated method stub
    super.onResume();
```

```
        music.paly(this, R.raw.yue);
    }
    @Override
    protected void onStop() {
        // TODO Auto-generated method stub
        super.onStop();
        music.stop(this);
    }
    @Override
    protected void onPause() {
        // TODO Auto-generated method stub
        super.onPause();
        music.stop(this);
    }
```

4. 配置 AndroidManifest.xml 文件

在 AndroidManifest.xml 当中添加 Prefs，代码如下。

```
<activity android:name="com.example.sudoku.Prefs"
    android:label="音效设置">
</activity>
```

任务拓展

为了加深对 PreferenceScreen 的了解，下面介绍一个 PreferenceScreen 的应用示例。实现效果如图 3-8 所示。

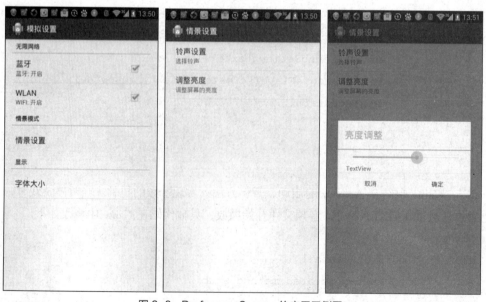

图 3-8　PreferenceScreen 的应用示例图

①设置 PreferenceScreen 的 xml 文件，根元素为 PreferenceScreen 代表显示整个屏幕；内部嵌套 PreferenceCategory 标签，表示偏好类别。在 PreferenceCategory 标签内部可以随便存放复选框、输入框、列表等显示控件。

Prefs.xml 代码如下。

```xml
<?xml version="1.0"encoding="utf-8"?>
<PreferenceScreen
xmlns:android="http://schemas.android.com/apk/res/android">
    <PreferenceCategory android:title="无限网络">
        <CheckBoxPreference
        android:defaultValue="true"
        android:key="Blueid"
        android:summaryOff="蓝牙：关闭"
        android:summaryOn="蓝牙：开启"
        android:title="蓝牙"/>
    <CheckBoxPreference
        android:defaultValue="true"
        android:key="wlanid"
        android:summaryOff="WIFI：关闭"
        android:summaryOn="WIFI：开启"
        android:title="WLAN"/>
</PreferenceCategory>
<PreferenceCategory android:title="情景模式">
<PreferenceScreen android:title="情景设置">
    <RingtonePreference
    android:key="选择铃声"
    android:summary="选择铃声"
    android:title="铃声设置"/>
<com.example.test.SeekBarPreference
    android:dialogLayout="@layout/seekbar"
    android:dialogTitle="亮度调整"
    android:key="light"
    android:summary="调整屏幕的亮度"
    android:title="调整亮度">
</com.example.test.SeekBarPreference>
</PreferenceScreen>
</PreferenceCategory>
```

```xml
<PreferenceCategory android:title="显示">
    <ListPreference
    android:dialogTitle="ListPreference"
    android:entries="@array/entries_list_preference"
    android:entryValues="@array/entriesvalue_list_preference"
    android:key="ListPreference"
    android:title="字体大小"/>
    </PreferenceCategory>
</PreferenceScreen>
```

② 创建类继承自 PreferenceFragment，通过 addPreferencesFromResource 方法将该 xml 文件加载到显示页面。

```java
public class Fragment extends PreferenceFragment{
    @Override
    public void onCreate(Bundle savedInstanceState) {
        // TODO Auto-generated method stub
        super.onCreate(savedInstanceState);
        addPreferencesFromResource(R.xml.prefs);
    }
}
```

③ 创建类继承自 Activity，利用 FragmentManager 对 PreferenceFragment 进行控制。该 Activity 类的 onCreate(Bundle savedInstanceState)内容如下。

```java
@Override
protected void onCreate(Bundle savedInstanceState) {
    super.onCreate(savedInstanceState);
    setContentView(R.layout.activity_main);
    FragmentManager fragmentManager = getFragmentManager();
    FragmentTransaction fragmentTransaction =
        fragmentManager.beginTransaction();
    Fragment fragment1 = new Fragment();
    fragmentTransaction.replace(android.R.id.content, fragment1);
    fragmentTransaction.addToBackStack(null);
    fragmentTransaction.commit();
}
```

④ 在本拓展任务中，亮度设置使用的是自定义 SeekBarPreference 设置。实现该功能需要两个步骤：设置 seekbar.xml 和定义 SeekBarPreference 类，内容如下。

注：SeekBar 的用法适用于 SDK4.1，4.2 版本不适用。

```xml
<?xml version="1.0"encoding="utf-8"?>
<LinearLayout xmlns:android="http://schemas.android.com/apk/res/android"
    android:layout_width="fill_parent"
    android:layout_height="fill_parent"
    android:orientation="vertical">
<SeekBar
    android:id="@+id/seekBar1"
    android:layout_width="fill_parent"
    android:layout_height="wrap_content"
    android:layout_marginLeft="20dip"
    android:layout_marginRight="10dip"
    android:max="100"
    android:progress="60">
</SeekBar>
<TextView
    android:id="@+id/textView1"
    android:layout_width="fill_parent"
    android:layout_height="wrap_content"
    android:layout_marginLeft="20dip"
    android:text="TextView">
</TextView>
</LinearLayout>
```

⑤ 定义 SeekBarPreference 类。

```java
import android.content.Context;
import android.preference.DialogPreference;
import android.util.AttributeSet;
import android.util.Log;
import android.view.View;
import android.widget.SeekBar;
import android.widget.TextView;
import android.widget.SeekBar.OnSeekBarChangeListener;
public class SeekBarPreference extends DialogPreference implements OnSeekBarChangeListener {
    private SeekBar seekBar;
    private TextView textView;
    public SeekBarPreference(Context context, AttributeSet attrs) {
        super(context, attrs);
```

```java
            // TODO Auto-generated constructor stub
        }
        protected void onBindDialogView(View view) {
            // TODO Auto-generated method stub
            super.onBindDialogView(view);
            seekBar = (SeekBar) view.findViewById(R.id.seekBar1);
            textView = (TextView) view.findViewById(R.id.textView1);
            seekBar.setOnSeekBarChangeListener(this);
        }
        protected void onDialogClosed(boolean positiveResult) {
            // TODO Auto-generated method stub
            if (positiveResult) {
                Log.i("Dialog closed", "You click positive button");
            } else {
                Log.i("Dialog closed", "You click negative button");
            }
        }
        public void onProgressChanged(SeekBar seekBar, int progress,boolean fromUser) {
            textView.setText(progress + "% " + progress + "/100");
        }
        public void onStartTrackingTouch(SeekBar seekBar) {
            // TODO Auto-generated method stub
        }
        @Override
        public void onStopTrackingTouch(SeekBar seekBar) {
            // TODO Auto-generated method stub
        }
    }
```

任务 3.5 继续游戏功能实现

任务描述

游戏中途停止后，可以通过"继续玩游戏"进入上一次游戏的数据界面，并在游戏开始时实现难度选择，效果如图 3-9 所示。

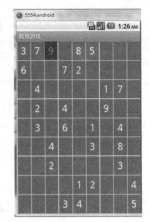

图 3-9 继续游戏效果图

任务目标

① 理解 Activity 的生命周期的概念。

② 结合 Activity 的 onCreate()、onPause()、onResume()、onStop()方法实现数据的存入与读取。

③ 掌握 SharedPreferences 的文件的作用。

④ 利用 SharedPreferences 文件实现数据的存取。

任务分析

本任务具体实现过程：

① 在 MainActivity 中设置"开始游戏"的难度和"继续游戏"的参数；

② Game 和 PuzzleView 中进行参数接收与处理；

③ 实现游戏暂停和继续功能。

知识要点

Activity 的生命周期

Activity 是 Android 程序与用户交流的桥梁，程序员通过 Activity 可以将前端服务，如 View 布局等各种事件消息展现给用户，是 Android 最基本的模块之一。一个 Activity 通常展现为一个可视化的用户界面，在一个用户界面中的 Activity 可以调用其他的 Activity 来启动另一个用户界面。多个独立 Activity 之间联系需要 Intent 来传值，一个 Android 应用程序至少有一个 Activity。通过调用 Activity 的 setContentView()方法来设置展现 Activity 的窗口的视图。之后可以实例控件 Button、EditView、TextView 等或事件来设计自己的用户界面，如图 3-10 所示。

Activity 有自己的生命周期。当 Activity 从一种状态转变到另一种状态时，可调用以下保护方法来通知这种变化。

OnCreate()：大部分情况都要重写这个方法。创建的时候会调用这个方法。通过设置这个

Activity 设置布局控件可初始化数据等。

OnStart()：当控件被看到的时候，就调用 OnStart()方法在 OnCreate()之后或者 OnStop()后调用在回到某个界面。

OnResume()：当重新获得用户焦点的时候就需调用。

onPause()：保留当前界面的数据，暂时地失去焦点，可能被另外一个透明的 Dialog 窗口覆盖，但是仍然和窗口管理器保持连接，系统可以继续保护 Activity 的一切活动。

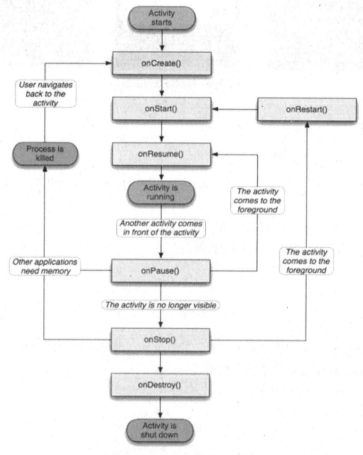

图 3-10　Activity 运行过程图

onStop()停止和 OnStart()对应注意 OnStop()和 OnPause()不一样的地方是 onStop( )完全被另外一个窗口覆灭，系统不能继续保持 Activity 的内部状态。

onDestory()销毁，到此这个 Activity 终于停止。

注意：在覆盖以上 7 个方法时，需要调用 super 不能覆盖了它本身的功能。

任务实现

1．设置"开始游戏"的难度和"继续游戏"的参数

在 MainActivity 中设置"开始游戏"的难度和"继续游戏"的参数。根据开始游戏时选择的难度等级或继续游戏的参数来确定启动游戏界面初识数据的难度，所以要在 MainActivity

中 startNewGame()方法中添加参数 diff。继续游戏参数 diff 值是-1，开始游戏时参数 diff 值是根据在对话框的难度选择来确定，难度等级有较难、中等、容易，对应的参数值为 0、1、2。

MainActivity 修改前的代码如下。

```
beginGame.setOnClickListener(this);
//事件处理
case R.id.gamebeging:
    startNewGame();
    break;
……
private void startNewGame(){
    Intent intent=new Intent(this,Game.class);
    startActivity(intent);
    }
……
```

MainActivity 修改后的代码。

```
//事件监听
beginGame.setOnClickListener(this);
continueGame.setOnClickListener(this);
//事件处理
case R.id.gamecontinue:
startGame(-1);
        break;
    case R.id.gamebegin:
        openNewGame();
        break;
//进入游戏界面
private void startGame(int diff) {
        // TODO Auto-generated method stub
        Intent intent=new Intent(this,Game.class);
        intent.putExtra("difficulty", diff);
        startActivity(intent);
    }
//开始新游戏
    private void openNewGame(){
        AlertDialog.Builder new_game=new AlertDialog.Builder(this);
        final String ss[]={"较难","中等","容易"};
        new_game.setTitle("选择难度");
        new_game.setItems(ss, new DialogInterface.OnClickListener() {
```

```
        @Override
                public void onClick(DialogInterface dialog, int which) {
                    Toast.makeText(MainActivity.this, ss[which], Toast.LENGTH_SHORT).show();
                    startGame(which);
                }
            });
            AlertDialog alert=new_game.create();
            alert.show();
        }
```

2. Game 和 PuzzleView 类中的参数接收与处理

开始游戏和继续游戏时通过参数传递进行区分的，而 Game 类中创建 PuzzleView 对象时需要传递参数给 PuzzleView，所以 PuzzleView 构造方法也需要添加参数，并根据参数进行相应的处理。

Game 类中的 onCreate 方法创建 PuzzleView 代码修改如下。

```
int diff=getIntent().getIntExtra("difficulty", 0);
puzzleview=new PuzzleView(this,diff);
```

PuzzleView 类中需定义 3 种难度的二维数组，并在 PuzzleView 的构造方法进行匹配，具体代码修改如下。

```
public String[][]  puzzle = new String[9][9];
    private String easy[][] = new String[][]{
                {"3","6"," "," "," "," "," "," "," "},
                {" "," ","4","2","3"," "," "," "," "},
                {" "," "," "," "," "," ","4","2"," "," "},
                {" ","7"," ","4","6"," "," "," ","3"},
                {"8","2"," "," "," "," "," ","1","4"},
                {"5"," "," "," ","1","3"," ","2"," "},
                {" "," ","1","9"," "," "," "," "," "},
                {" "," ","7"," ","4","8","3"," "," "},
                {" "," "," "," "," "," "," ","4","5"},
        };
        private String medium[][] = new String[][]{
                {"6","5"," "," "," "," "," ","7"," "},
                {" "," "," "," ","5"," "," ","6"," "," "," "},
                {" ","1","4"," "," "," "," "," ","5"},
                {" ","7"," ","4","6"," "," "," ","3"},
                {" "," ","2","3","1","4","7"," "," "},
                {" "," "," ","7"," "," "," ","8"," "," "},
```

```java
        {"5"," "," "," "," "," ","6","3"," "},
        {" ","9"," ","3"," ","1"," ","8"," "},
        {" "," "," "," "," "," ","6"," "," "},
};
private String hard[][] = new String[][]{
        {"1","2","3","4","5","6","7","8","9"},
        {" "," "," "," "," "," "," "," "," "},
        {" "," "," "," "," "," "," "," "," "},
        {" "," "," "," "," "," "," "," "," "},
        {" "," "," "," "," "," "," "," "," "},
        {" "," "," "," "," "," "," "," "," "},
        {" "," "," "," "," "," "," "," "," "},
        {" "," "," "," "," "," "," "," "," "},
        {" "," "," "," "," "," "," "," "," "},
};
public PuzzleView(Context context, int diff) {
    super(context);
    new_game=(Game)context;
    setFocusable(true);//允许键盘事件为true,才会响应该事件
    setFocusableInTouchMode(true);//允许触屏事件为true才会响应相应事件
    for(i=0;i<9;i++)
        for(j=0;j<9;j++)
        {
            switch(diff)
            {
            case 0:
                puzzle[i][j] = easy[i][j];
                break;
            case 1:
                puzzle[i][j] = medium[i][j];
                break;
            case 2:
                puzzle[i][j] = hard[i][j];
                break;
            case -1:
                puzzle[i][j] = new_game.continueString.substring(m, m+1);
                System.out.print(puzzle[i][j]);
```

```
                    m++;
                    break;
            }
        }
    }
```

3. MainActivity 中实现游戏暂停和继续功能

在 onCreate()方法中,从 SharedPreferences 文件获取 puzzle 内容。

```
continueString = getPreferences(MODE_PRIVATE).getString("puzzle", "");
```

在 Game 中重写 Activity 生命周期的方法 onResume()、onPause(),并添加 arraytoString() 方法处理字符串。

① 在 onPause()方法中游戏暂停时将游戏界面的数组利用 edit().putString()方法存放到 SharedPreferences 文件中。

② 在 onResume()游戏重新唤醒时通过 SharedPreferences 的 getString 方法获取 SharedPreferences 当中的数据。

③ arraytoString()方法实现将二维数组添加到字符串当中。

```
@Override
    protected void onResume() {
        // TODO Auto-generated method stub
        super.onResume();
        continueString=getPreferences(MODE_PRIVATE).getString("puzzle", "");
        music.paly(this, R.raw.yue);
    }
    @Override
    protected void onPause() {
        // TODO Auto-generated method stub
        super.onPause();
        getPreferences(MODE_PRIVATE).edit().putString("puzzle",
arraytoString()).commit();
        music.stop(this);
    }
/*
    * 把 puzzle 数组中的 9*9 字符串
    * */
    private String arraytoString(){
        String s1="";
        int j,k;
        for(j=0;j<9;j++)
            for(k=0;k<9;k++){
```

```
            s1+=puzzleview.puzzle[j][k];
        }
    return s1;
}
```

项目四 手机定位应用

PART 4

项目情境

手机定位应用适用于手机丢失、被窃或携带手机的儿童、智障人士、老人的找寻。通过 GPS 全球定位系统（Global Positioning System，GPS），向丢失手机发送指令，丢失手机返回该手机所在位置的信息。在备用手机的百度地图中显示丢失手机的地理位置。该应用程序主要包含短信发送、接收、解析和 GPS 定位功能，实现对联系人的手机定位，如图 4-1 所示。

图 4-1　手机定位运行效果

学习目标

☑ 巩固界面设计、事件处理、SQLite 数据管理。
☑ 理解短信的接收与拦截原理，掌握短信的接收与拦截实现方法。

☑ 理解 GPS 定位实现原理，掌握 GPS 定位实现方法。
☑ 理解地图编程的原理，掌握地图应用编程基本方法。

工作任务

任务名称
任务 4.1　界面与数据层设计
任务 4.2　短信发送与接收处理
任务 4.3　地图显示联系人位置

任务 4.1　界面与数据层设计

任务描述

创建登录界面如图 4-2 左所示，验证登录用户名和密码与数据库中的信息是否一致，实现连接数据库的用户登录验证。

创建界面如图 4-2 右所示，其中"设置"按钮实现将设置的数据进行保存。"查看联系人位置"按钮实现将查看信息发送到联系人，"备份联系人信息至服务器"按钮则将备份信息发送到联系人，"备份联系人图片至服务器"则发送备份消息至联系人手机上。

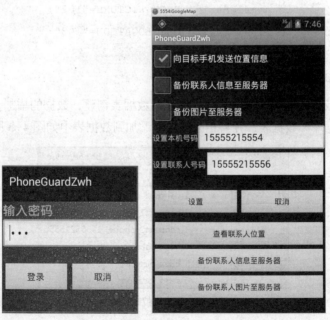

图 4-2　手机定位登录和主界面效果

任务目标

① 通过任务巩固界面设计，如布局与常用控件的设置。
② 巩固 SQLite 数据库应用编程。

任务分析

① 设计并实现持久层。
② 设计并实现数据层。
③ 实现“用户登录”功能。
④ 实现“参数设置”功能。

任务实现

1. 设计并实现持久层

在系统中常常有一些数据不会发生变化，但在各个模块中都需要用到这些信息，因此将这些不发生变化的数据设计在持久层 CommonUtils 中，供其他类使用。

```
public final class CommonUtils {
    private CommonUtils(){}
    public static final String PROTOCOL="protocol";
    publi cstatic final String CMD="cmd";
    public static final String SEND_LOCATION="100";
    public static final String BACK_CONTRACT="101";
    public static final String BACK_PICTURE="102";
    public static final String GET_LOCATION="103";
}
```

2. 设计并实现数据层

设计并实现数据处理层，实现数据库创建，数据表创建，数据的增加、删除、修改和查询方法。创建数据库和数据表，并将相应的内容添加到数据表中如图 4-3 所示。

RecNo	_id	phone_key	phone_value
Click here to define a filter			
1	1	col_pwd	asd
2	2	contract1_phone	15555215554
3	3	contract2_phone	15555215556
4	4	back_contract	0
5	5	back_picture	0
6	6	send_location	1

图 4-3　手机定位数据内容

① 创建 PhoneDB 类，该类继承自 SQLiteOpenHelper，通过该类的构造函数创建数据库

PhoneDB。

② 在 onCreate()方法中使用 execSQL()方法实现数据表的创建和数据的插入。该表的结构和数据表的初始内容如图 4-3 所示。

③ 创建 public String query(String key)方法，该方法通过 SQLiteOpenHelper 类的 get ReadableDatabase()方法实现数据库的创建，根据输入的 key 值查询 phone_key=key 的 phone_value 值，并返回该行数据的 phone_value 值。

④ 创建 public boolean update(String key,String value)方法，该方法根据输入的 key 和 value 值，利用数据库的 update 方法实现数据库的更新。

```java
package com.example.phoneguardzwh;
import android.content.ContentValues;
import android.content.Context;
import android.database.Cursor;
import android.database.sqlite.SQLiteDatabase;
import android.database.sqlite.SQLiteOpenHelper;
import android.util.Log;
public class PhoneDB extends SQLiteOpenHelper {
    private static final String TABLE_NAME="settings";
    public static final String PHONE_KEY="phone_key";
    public static final String PHONE_VALUE="phone_value";
    public static final String COL_ID="_id";
    public static final String COL_PWD="col_pwd";
    public static final String COL_CONTRACT1_PHONE="contract1_phone";
    public static final String COL_CONTRACT2_PHONE="contract2_phone";
    public static final String COL_SEND_LOCATION="send_location";
    public static final String COL_BACK_CONTRACT="back_contract";
    public static final String COL_BACK_PICTURE="back_picture";
    public PhoneDB(Context context){
        super(context, "phone.db", null, 1);
    }
    @Override
    public void onCreate(SQLiteDatabase db) {
        Log.i("info","PhoneDB-onCreate()");
        String sql="create table "+TABLE_NAME+"("+
            COL_ID+" integer primary key autoincrement,"+
            PHONE_KEY+" text not null,"+
            PHONE_VALUE+" text not null)";
        db.execSQL(sql);
        sql="insert into "+TABLE_NAME+
```

```java
            "("+PHONE_KEY+","+PHONE_VALUE+")values("+
            "'"+COL_PWD+"','asd')";
    db.execSQL(sql);
    sql="insert into "+TABLE_NAME+
            "("+PHONE_KEY+","+PHONE_VALUE+")values("+
            "'"+COL_CONTRACT1_PHONE+"','15555215554')";
    db.execSQL(sql);
    sql="insert into "+TABLE_NAME+
            "("+PHONE_KEY+","+PHONE_VALUE+")values("+
            "'"+COL_CONTRACT2_PHONE+"','15555215556')";
    db.execSQL(sql);
    sql="insert into "+TABLE_NAME+
            "("+PHONE_KEY+","+PHONE_VALUE+")values("+
            "'"+COL_BACK_CONTRACT+"','0')";
    db.execSQL(sql);
    sql="insert into "+TABLE_NAME+
            "("+PHONE_KEY+","+PHONE_VALUE+")values("+
            "'"+COL_BACK_PICTURE+"','0')";
    db.execSQL(sql);
    sql="insert into "+TABLE_NAME+
            "("+PHONE_KEY+","+PHONE_VALUE+")values("+
            "'"+COL_SEND_LOCATION+"','1')";
    db.execSQL(sql);
}
@Override
public void onUpgrade(SQLiteDatabase db, int oldVersion, int newVersion) {
}
public String query(String key){
    SQLiteDatabase db=getReadableDatabase();
    String[] selection={COL_ID,PHONE_KEY,PHONE_VALUE};
    Cursor cursor=db.query(TABLE_NAME, selection,
            PHONE_KEY+"='"+key+"'", null, null, null,null);
    if(cursor==null){
        return null;
    }
    if(cursor.moveToFirst()){
        int col=cursor.getColumnIndexOrThrow(PHONE_VALUE);
        return cursor.getString(col);
```

```
            }
            return null;
        }
    public boolean update(String key,String value){
        SQLiteDatabase db=getWritableDatabase();
        ContentValues values=new ContentValues();
        values.put(PHONE_VALUE, value);
        return db.update(TABLE_NAME, values, PHONE_KEY+"='"+key+"'", null)>0;
        }
    }
```

3．实现"用户登录"功能

设计登录界面并实现登录功能，设计登录界面 login.xml，该界面控件如图 4-2 所示。创建 Activity 类 Login.java 将其显示界面设置为 login.xml，并加载 login.xml 界面中的各个控件。在登录按钮单击事件通过调用数据层的 query()方法实现登录验证，验证成功后使用 Intent。

```
package com.example.phoneguardzwh;
import android.app.Activity;
import android.content.Intent;
import android.os.Bundle;
import android.util.Log;
import android.view.View;
import android.view.View.OnClickListener;
import android.widget.Button;
import android.widget.EditText;
public class LoginActivity extends Activity implements OnClickListener{
    EditText metPwd;
    Button mbtnLogin,mbtnExit;
    @Override
    public void onCreate(Bundle savedInstanceState) {
        super.onCreate(savedInstanceState);
        setContentView(R.layout.login);
        PhoneDB phoneDb=new PhoneDB(this);
        String pwd=phoneDb.query(PhoneDB.COL_PWD);
        Log.i("info",pwd);
        boolean b=phoneDb.update(PhoneDB.COL_BACK_CONTRACT, "1");
        if(b){
     Log.i("info",phoneDb.query(PhoneDB.COL_BACK_CONTRACT));
        }
```

```java
        initViews();
    }
    private void initViews() {
        metPwd=(EditText) findViewById(R.id.etPwd);
        mbtnExit=(Button) findViewById(R.id.btnExit);
        mbtnLogin=(Button) findViewById(R.id.btnLogin);
        mbtnExit.setOnClickListener(this);
        mbtnLogin.setOnClickListener(this);
    }
    @Override
    public void onClick(View v) {
        switch (v.getId()) {
        case R.id.btnExit:
            finish();
            break;
        case R.id.btnLogin:
            String pwd=metPwd.getText().toString();
            if(pwd==null||pwd.length()==0){
                metPwd.setError("请输入密码");
                return ;
            }
            PhoneDB Phonedb=new PhoneDB(this);
            String password=Phonedb.query(PhoneDB.COL_PWD);
            if(!pwd.trim().equals(password)){
                metPwd.setError("密码错误");
            }
            Intent intent=new Intent(this, SettingsActivity.class);
            startActivity(intent);
            finish();
            break;
        }
    }
}
```

4．实现"参数设置"功能

设计并实现参数设置界面，包括设置参数、查看联系人、备份联系人、备份联系人图片至服务器等事件响应。

"设置"功能的界面设计通过 settings.xml 文件实现，创建 Activity 类 SettingsActivity 用于

实现设置的各项功能。SettingsActivity 的各项方法内容介绍如下。

① initViews ()方法用于将 settings.xml 界面的各个控件加载到 SettingsActivity 当中。

② initCheckBoxs()方法，首先查询数据库 PhoneDB.COL_SEND_LOCATION、PhoneDB.COL_BACK_CONTRACT、PhoneDB.COL_BACK_PICTURE 的值，然后根据查询的值利用 setChecked()方法进行复选框的设置。

③ private void update()方法，调用 PhoneDB 的 update 方法实现数据的更新。更新的数据方法如下：phonedb.update(PhoneDB.COL_BACK_CONTRACT, mchkBackContract.isChecked()?"1":"0")。

依此方法更新 PhoneDB.COL_BACK_PICTURE、PhoneDB.COL_SEND_LOCATION 的值。从文本框中获取本机和联系人的号码，并使用 phonedb.update()方法对号码进行更新。

```java
package com.example.phoneguardzwh;
import java.util.List;
import android.app.Activity;
import android.app.PendingIntent;
import android.content.Intent;
import android.os.Bundle;
import android.telephony.SmsManager;
import android.view.View;
import android.view.View.OnClickListener;
import android.widget.Button;
import android.widget.CheckBox;
import android.widget.EditText;
public class SettingsActivity extends Activity implements OnClickListener{
    CheckBox mchkSendLocation;//设置接收联系人要求发送本机位置
    CheckBox mchkBackContract;//设置接收联系人要求备份本机联系人至服务器
    CheckBox mchkBackPicture;//设置接收联系人要求备份本机图片至服务器
    Button mbtnOk,mbtnCancel;//设置、取消按钮
    Button mbtnSendLocation;//发送查看联系人位置短信的按钮
    Button mbtnBackContract;//发送备份联系人至服务器的短信的按钮
    Button mbtnBackPicture;//发送备份联系人图片至服务器的短信的按钮
    EditText metContract1Phone;//存放本机号码
    EditText metContract2Phone;//设置联系人号码
    @Override
    protected void onCreate(Bundle savedInstanceState) {
        super.onCreate(savedInstanceState);
        setContentView(R.layout.settings);
        initViews();
        initCheckBoxs();
```

```java
    }
    private void initCheckBoxs() {
        PhoneDB phonedb=new PhoneDB(this);
        String sendLocation=phonedb.query(PhoneDB.COL_SEND_LOCATION);
        String backContract=phonedb.query(PhoneDB.COL_BACK_CONTRACT);
        String backPicture=phonedb.query(PhoneDB.COL_BACK_PICTURE);
        mchkBackContract.setChecked(backContract.equals("1"));
        mchkBackPicture.setChecked(backPicture.equals("1"));
        mchkSendLocation.setChecked(sendLocation.equals("1"));
        metContract1Phone.setText(phonedb.query(PhoneDB.COL_CONTRACT1_PHONE));
        metContract2Phone.setText(phonedb.query(PhoneDB.COL_CONTRACT2_PHONE));
    }
    private void initViews() {
        mchkBackContract=(CheckBox) findViewById(
            R.id.chkBackContract);
        mchkBackPicture=(CheckBox) findViewById(
            R.id.chkBackPicture);
        mchkSendLocation=(CheckBox) findViewById(
            R.id.chkSendLocation);
        mbtnCancel=(Button) findViewById(R.id.btnCancel);
        mbtnOk=(Button) findViewById(R.id.btnOk);
        mbtnCancel.setOnClickListener(this);
        mbtnOk.setOnClickListener(this);
        metContract1Phone=(EditText) findViewById(
            R.id.etContract1Phone);
        metContract2Phone=(EditText) findViewById(
            R.id.etContract2Phone);
        mbtnSendLocation=(Button) findViewById(
            R.id.btnSendLocation);
        mbtnBackContract=(Button) findViewById(
            R.id.btnBackContract);
        mbtnBackPicture=(Button) findViewById(
            R.id.btnBackPicture);
        mbtnSendLocation.setOnClickListener(this);
        mbtnBackContract.setOnClickListener(this);
        mbtnBackPicture.setOnClickListener(this);
    }
    @Override
```

```java
public void onClick(View v) {
    switch (v.getId()) {
    case R.id.btnCancel:
        finish();
        break;
    case R.id.btnOk:
        update();//根据设置更新数据库
        break;
    case R.id.btnSendLocation:
        sendSms(CommonUtils.SEND_LOCATION);
        break;
    case R.id.btnBackContract:
        sendSms(CommonUtils.BACK_CONTRACT);
        break;
    case R.id.btnBackPicture:
        sendSms(CommonUtils.BACK_PICTURE);
        break;
    }
}
//向联系人手机发送指定命令的短信
private void sendSms(String cmdType) {
    //该方法下一个任务中实现
}
//根据设置更新数据库
private void update() {
    PhoneDB phonedb=new PhoneDB(this);
    phonedb.update(PhoneDB.COL_BACK_CONTRACT, mchkBackContract.isChecked()?"1":"0");
    phonedb.update(PhoneDB.COL_BACK_PICTURE, mchkBackPicture.isChecked()?"1":"0");
    phonedb.update(PhoneDB.COL_SEND_LOCATION, mchkSendLocation.isChecked()?"1":"0");
    String contract1Phone=metContract1Phone.getText().toString();
    if(contract1Phone!=null&&contract1Phone.length()>0){
        phonedb.update(PhoneDB.COL_CONTRACT1_PHONE, contract1Phone);
    }else{
        metContract1Phone.setError("请输入手机号码");
    }
```

```
            String contract2Phone=metContract2Phone.getText().toString();
            if(contract2Phone!=null&&contract2Phone.length()>0){
                phonedb.update(PhoneDB.COL_CONTRACT2_PHONE, contract2Phone);
            }else{
                metContract2Phone.setError("请输入手机号码");
            }
        }
    }
}
```

技能训练

结合项目二介绍的网络知识将备份联系人、备份联系人图片的功能通过连接 Web 服务器端，实现网络数据备份功能。

任务 4.2 短信发送与接收处理

任务描述

创建消息接收处理类 RemoteSmsReceiver，根据短信内容进行相应的处理。该类实现查看联系人功能。

创建 SendSmsService 类实现联系人的 GPS 功能，并将定位的经纬度以短信的方式发送。

任务目标

① 熟练掌握短信发送的实现方法。
② 掌握短信接收与处理的方法。
③ 熟练掌握 GPS 定位的实现方法。
④ 通过完成任务功能锻炼逻辑思维能力。

任务分析

本任务要实现短信发送和接收处理功能，其业务过程如图 4-4 所示。

具体实现步骤分析如下。
① 实现短信的接收与处理功能。
② 发送"查看联系人"短信。
③ 发送 GPS 定位信息到联系人手机上。
④ 配置 AndroidManifest.xml 文件。

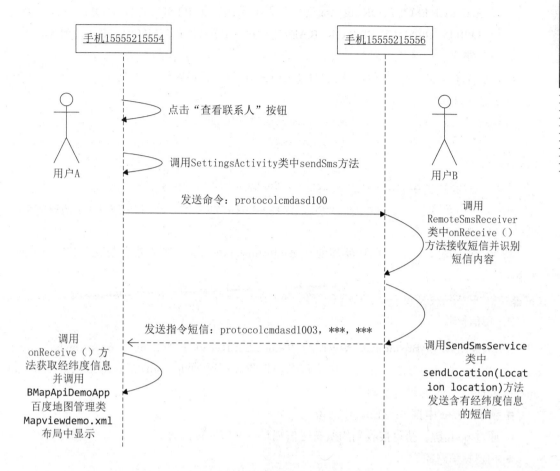

图 4-4　手机定位活动图

知识要点

该任务的实现主要涉及短信发送与接收、GPS 定位等相关知识。下面介绍短信发送与接收、GPS 定位的基本概念和常用方法。

1．短信发送

短信发送与接收需要使用短信管理类 SmsManager，通过该类创建短信管理对象实现短信发送。实现短信发送的步骤如下。

① 创建类 SmsManager 对象。smsManager 为 SmsManager 一个默认的实例。

② PendingIntent.getBroadcast(SMSDemo.this, 0, new Intent(), 0);相当于 Context.sendBroadcast();

③ sendTextMessage()方法的参数说明如下。

- destinationAddress: 收件人号码。
- scAddress: 短信中心服务号码，这里设置为 null。
- text: 发送内容。
- sentIntent: 如果不为空，当消息成功发送或失败这个 PendingIntent 就广播。结果代码是 Activity.RESULT_OK 表示成功，如果结果代码是 RESULT_ERROR_GENERIC_FAILURE、RESULT_ERROR_RADIO_OFF、RESULT_ERROR_NULL_PDU 之一表示错误。
- deliveryIntent：对方接收状态信号(是否已成功接收)，设置为 null 即可。

下面是实现短信发送的主要代码。

```
SmsManager smsManager = SmsManager.getDefault();
PendingIntent pintent = PendingIntent.getBroadcast(SMSDemo.this, 0, new Intent(), 0);
SmsManager.sendTextMessage(String destinationAddres s,
 String scAddress, String text, PendingIntent sentIntent, PendingIntent deliveryIntent)
```

需要注意的是，发送信息功能需要在 AndroidManifest.xml 加入短信发送权限。其代码如下。

```
<uses-permission android: name="android.permisson.SEND_SMS" />
```

2．短信接收

① 创建继承自 BroadcastReceiver 的 RemoteSmsReceiver 类，重写 onReceive(Context context, Intent intent)方法。在方法中实现短信接收和拦截功能，onReceive(Context context, Intent intent)方法的主要内容如下。

- 获取 intent 中携带的 Bundle 对象。
- 通过 pdus 键，获得接收到的所有短信消息，获取短信内容。
- 构建短信对象数组。
- 获取单条短信内容，以 pdu 格式存放，并生成短信对象。
- 打印发送短信的手机号和短信内容。
- 终止系统的短信广播。

要侦听传入的 SMS 消息，需要创建一个 BroadcastReceiver 类，该类通过 onReceive()方法使应用程序接受其他应用程序使用 sendBroadcast()发送的 Intent。onReceive()方法在收到一个传入的 SMS 消息时被触发。SMS 消息通过一个 Bundle 对象包含在 Intent 对象中进行传输。注意每收到一每 SMS 消息都会调用 onReceive()方法，如果设备收到多条 SMS 消息，则会多次调用 onReceive()方法。

每条 SMS 消息是以 PDU 格式存储在一个 Object 数组中。如果 SMS 消息少于 160 个字符，那么数组只包含一个元素。否则，该消息将会被分割成多条更小的消息，作为多个元素存储在数组中。

createFromPdu0 方法可以提取每条消息的内容。
getDisplayOriginatingAddress()方法用于获取消息中发件人电话号码。
getDisplayMessageBody()方法用于获取消息的正文。

```
public class RemoteSmsReceiver extends BroadcastReceiver {
    @Override
    public void onReceive(Context context, Intent intent) {
```

```java
//步骤1、获取intent中携带的Bundle对象
    Bundle bundle=intent.getExtras();
    if(bundle==null){
        return ;
    }
    //步骤2、通过pdus键，获得接收到的所有短信消息，获取短信内容；
    Object[] smsObject=(Object[]) bundle.get("pdus");
    StringBuilder sb=new StringBuilder();
    //步骤3、构建短信对象数组；
    SmsMessage[] messages=new SmsMessage[smsObject.length];
    for (int i = 0; i < messages.length; i++) {
    //步骤4、获取单条短信内容，以pdu格式存,并生成短信对象；
    messages[i]=SmsMessage.createFromPdu(
        (byte[])smsObject[i]);
    }
    //步骤5、打印发送短信的手机号和短信内容
    for(SmsMessage msg:messages){
        sb.append("短信来自")
        .append(msg.getDisplayOriginatingAddress())
        .append("\n"+msg.getDisplayMessageBody());
    }
    Log.i("info",sb.toString());
    //步骤6、终止系统的短信广播
    abortBroadcast();
}
```

② 注册广播接收器。

```xml
<receiver android:name="…">         提示：此处应写实际的短信接收类的包名和类名
    <intent-filterandroid:priority="1000">
    <action android:name="android.provider.Telephony.SMS_RECEIVED" />
    </intent-filter>
</receiver>
```

3. GPS定位

GPS定位需要用到LocationManager类，该类用于创建所有与GPS相关的类。
创建LocationManager对象，需要通过Context.getSystemService方法，示例代码如下。
LocationManager lm=Context.getSystemService(Context.LOCATION_SERVICE);
常用方法如下。
- List<String> getAllProvider()

作用：获取所有 LocationProvider 的名称。

提示：LocationProvider 类，管理定位信息。稍后详细介绍该类。

- String getBestProvider(Criteria criteria;boolean enableOnly)

作用：按指定条件获取最优的 LocationProvider 的名称。

说明：Criteria 类，负责设置选择 LocationProvider 的条件。

该类有以下常用方法。

SetAccuracy(int accuracy)

作用：设置对 LocationProvider 的精度要求。

setAltitudeRequired(Boolean altitudeRequired)

作用：设置对 LocationProvider 的高度要求。

setBearingRequired(Boolean bearingRequired);

作用：设置对 LocationProvider 的方向信息的要求。

- setCostAllowed(Boolean costAllowed)

作用：设置要求 LocationProvider 是否免费。

setPowerRequirement(int level)

作用：设置要求 LocationProvider 的耗电量。

setSpeedRequired(Boolean speedRequired)

作用：设置要求 LocationProvider 能提供的速度信息。

参数-enabledOnly：true：只从有效的 LocationProvider 中获取。

- List<String> getProviders(Criteria criteria;boolean enableOnly)

作用：获取所有符合指定条件的 LocationProvider 的名称。

- Location getLastKnownLocation(String provider)

作用：根据 LocationProvider 获取最近一次已知的 Location 对象。

说明：Location 对象管理位置信息，包括经度、维度值。

- LocationProvider getProvdier(String name)

作用：根据名称获取 LocationProvider 对象。

- List<String> getProviders(enabledOnly)

作用：返回所有有效的 LocationProvider 名称。

- void requestLocationUpdates(String provider,long minTime, float minDistance,Pending Intent intent)

作用：通过指定的 LocationProvider 周期性地获取定位信息，并通过 intent 启动目标组件。

参数-provider：指定的 LocationProvider。

参数-minTime：获取定位信息的时间间隔。

参数-minDistance：获取定位信息的最短距离。

- void requestLocationUpdates(String provider,long minTime,float minDistance, LocationListener listener)

作用：通过指定的 LocationProvider，按指定时间间隔、指定最短距离的条件获取定位信息，并触发 LocationListener 监听器事件。

参数-provider：指定的 LocationProvider。

参数-minTime：获取定位信息的时间间隔。

参数-minDistance：获取定位信息的最短距离。

参数-listener：触发的 LocationListener 监听事件。

说明：LocationListener 接口负责处理当位置发生改变时触发的事件。

常用的常量如下。

GPS_PROVIDER：用于创建 LocationProvider 对象。示例代码：
LocationProvider provider=LocationManager.GPS_PROVIDER;

KEY_PROXIMITY_ENTERING：键名，用于获取是否达到临近警告的状态值。示例：
Boolean isInto=intent.getBooleanExtra(
LocationManager.KEY_PROXIMITY_ENTERING);

任务实现

1. 实现短信的接收与处理功能

创建短信接收处理类 RemoteSmsReceiver，用于接收对方联系人的短信信息，分析短信内容，根据短信内容进行相应的处理。

短信的内容分析如下。

- 短信的 0~8 个字符为 CommonUtils.PROTOCOL 即 protocol。
- 8~11 字符内容为 CommonUtils.CMD，即 cmd。
- 短信 11 个字符以后的内容为密码，通过数据查找密码，此处为 asd。
- 密码内容之后是短信请求类型，100 为请求获取对方位置信息，103****，****为发送本机的经纬度信息给联系人。短信内容分析示例如下。

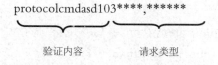

示例中的 protocolcmdasd 是联系人对短信验证内容，103****，****请求类型为 GET_LOCATION——获取联系人位置信息及其位置的精度纬度。

```
public class RemoteSmsReceiver extends BroadcastReceiver {
    @Override
    public void onReceive(Context context, Intent intent) {
        StringBuilder sb=new StringBuilder();
        Bundle bundle=intent.getExtras();
        if(bundle==null){
            return ;
        }
        Object[] smsObject=(Object[]) bundle.get("pdus");
```

```java
SmsMessage[] messages=new SmsMessage[smsObject.length];
for (int i = 0; i < messages.length; i++) {
    messages[i]=SmsMessage.createFromPdu((byte[])smsObject[i]);
}
String contract1Phone=query(context,PhoneDB.COL_CONTRACT1_PHONE);
String contract2Phone=query(context,PhoneDB.COL_CONTRACT2_PHONE);
String incomeNumber;
for(SmsMessage message:messages){
    incomeNumber=message.getDisplayOriginatingAddress();
    if(!incomeNumber.equals(contract1Phone)&&
        !incomeNumber.equals(contract2Phone)){
            continue;
    }
    sb.append(message.getDisplayMessageBody());
}
if(sb==null||sb.length()==0){
    return ;
}
if(!sb.substring(0,8).equals(CommonUtils.PROTOCOL)){
    return ;
}
if(!sb.substring(8,11).equals(CommonUtils.CMD)){
    return ;
}
String pwdDb=query(context, PhoneDB.COL_PWD);
int index=sb.indexOf(pwdDb);
if(index<11){
    return ;
}
//获取命令类型在短信中的起始位置
int cmdStart=11+pwdDb.length();
//获取命令类型的字符串
String cmdType=sb.substring(cmdStart,cmdStart+3);
//向联系人手机发送本机位置的短信
if(cmdType.equals(CommonUtils.SEND_LOCATION)){
    Log.i("info","查看联系人的信息");
    abortBroadcast();
    Intent service=new Intent(context, SendSmsService.class);
```

```
                context.startService(service);
            }
        }elseif(cmdType.equals(CommonUtils.GET_LOCATION)){
            Log.i("info","查看地图");
            //若是显示位置的命令
            abortBroadcast();
            //从短信中获取经纬度的字符串
            String strLocation=sb.substring(cmdStart+3);
            Intent intentAct=new Intent(context,
                LocationOverlay.class);
            intentAct.putExtra("location", strLocation);
            intentAct.setFlags(Intent.FLAG_ACTIVITY_NEW_TASK);
            //跳转至显示联系人位置的窗口
            context.startActivity(intentAct);
        }
    }
    //封装数据库查询操作
    public String query(Context context,String key){
        PhoneDB phonedb=new PhoneDB(context);
        return phonedb.query(key);
    }
}
```

2. 发送"查看联系人"短信指令

在"设置"功能的 SettingsActivity 中完成 sendSms(String cmdType)方法,实现向联系人手机发送指定命令的短信。

```
private void sendSms(String cmdType) {
    StringBuilder sb=new StringBuilder(
        CommonUtils.PROTOCOL);
    sb.append(CommonUtils.CMD);
    PhoneDB phonedb=new PhoneDB(this);
    String pwd=phonedb.query(PhoneDB.COL_PWD);
    sb.append(pwd);
    if(cmdType.equals(CommonUtils.SEND_LOCATION)){
        sb.append(CommonUtils.SEND_LOCATION);
    }elseif(cmdType.equals(CommonUtils.BACK_CONTRACT)){
        sb.append(CommonUtils.BACK_CONTRACT);
    }elseif(cmdType.equals(CommonUtils.BACK_PICTURE)){
        sb.append(CommonUtils.BACK_PICTURE);
```

```java
        }
        SmsManager smsManager=SmsManager.getDefault();
        List<String> messages=smsManager.divideMessage(
            sb.toString());
        String contract2Phone=phonedb.query(
            PhoneDB.COL_CONTRACT2_PHONE);
        PendingIntent pi=PendingIntent.getBroadcast(
            this, 0, new Intent(), 0);
        for(String message:messages){
            smsManager.sendTextMessage(contract2Phone, null,
                message, pi, null);
        }
    }
```

3. 发送 GPS 定位信息到联系人手机上

创建 SendSmsService 类，该类通过 LocationManager 实现位置的获取，并发送带有 GPS 地址的短信到对方联系人手机上。

```java
package com.example.phoneguardzwh;
import java.util.ArrayList;
import android.app.PendingIntent;
import android.app.Service;
import android.content.Intent;
import android.location.Location;
import android.location.LocationListener;
import android.location.LocationManager;
import android.os.Bundle;
import android.os.IBinder;
import android.telephony.SmsManager;
public class SendSmsService extends Service {
    LocationManager mLocationManager;
    @Override
    public IBinder onBind(Intent intent) {
        // TODO Auto-generated method stub
        return null;
    }
    @Override
    public int onStartCommand(Intent intent, int flags, int startId) {
        //实例化位置服务管理器对象
        mLocationManager=(LocationManager) getSystemService(LOCATION_SERVICE);
```

```java
        //设置位置改变的事件监听
        mLocationManager.requestLocationUpdates(LocationManager.GPS_PROVIDER,
30, 10, new LocationListener() {
            @Override
            public void onStatusChanged(String provider, int status,
Bundle extras) {
            }
            @Override
            public void onProviderEnabled(String provider) {
                Location location=mLocationManager.getLastKnownLocation(provider);
                sendLocation(location);
            }
            @Override
            public void onProviderDisabled(String provider) {
            }
            @Override
            public void onLocationChanged(Location location) {
                sendLocation(location);
            }
        });
        return super.onStartCommand(intent, flags, startId);
    }
    //向联系人发送本机位置的短信
    private void sendLocation(Location location) {
        double latitude=location.getLatitude();
        double longtitude=location.getLongitude();
        SmsManager smsManager=SmsManager.getDefault();
        //发送带有经纬度的协议的字符串
        String strLoation=CommonUtils.PROTOCOL+
            CommonUtils.CMD+
            new PhoneDB(this).query(PhoneDB.COL_PWD)+
            CommonUtils.GET_LOCATION+
            latitude+","+longtitude;
        PendingIntent pi=PendingIntent.getBroadcast(this, 0, new Intent(), 0);
        ArrayList<String> messages=smsManager.divideMessage(strLoation);
        String targetNumber=new PhoneDB(this).query(PhoneDB.COL_CONTRACT2_PHONE);
```

```
        for(String message:messages){
            smsManager.sendTextMessage(targetNumber, null,
                message, pi, null);
        }
    }
}
```

4. 配置 AndroidManifest.xml 文件

本任务中涉及短信发送与接收功能，因此需要在 AndroidManifest.xml 文件中注册相应的权限，并<application>元素下注册<receiver>和<service>元素。

权限列表内容如下。

```xml
<uses-permission android:name="android.permission.ACCESS_FINE_LOCATION" >
    </uses-permission>
    <uses-permission android:name="android.permission.READ_SMS" >
    </uses-permission>
    <uses-permission android:name="android.permission.SEND_SMS" >
    </uses-permission>
    <uses-permission android:name="android.permission.RECEIVE_SMS" >
    </uses-permission>
    <uses-permission android:name="android.permission.ACCESS_NETWORK_STATE" >
    </uses-permission>
    <uses-permission android:name="android.permission.ACCESS_FINE_LOCATION" >
    </uses-permission>
    <uses-permission android:name="android.permission.INTERNET" >
    </uses-permission>
    <uses-permission android:name="android.permission.WRITE_EXTERNAL_STORAGE" >
    </uses-permission>
    <uses-permission android:name="android.permission.ACCESS_WIFI_STATE" >
    </uses-permission>
    <uses-permission android:name="android.permission.CHANGE_WIFI_STATE" >
    </uses-permission>
    <uses-permission android:name="android.permission.READ_PHONE_STATE" >
    </uses-permission>
    <uses-permission android:name="android.permission.CALL_PHONE" >
</uses-permission>
```

以下内容在<application>元素中添加。

```xml
<receiver android:name=".RemoteSmsReceiver" >
    <intent-filter android:priority="1000" >
        <action android:name="android.provider.Telephony.SMS_RECEIVED" />
    </intent-filter>
</receiver>
<service android:name=".SendSmsService"/>
```

任务拓展

与数据库结合的防骚扰短信应用编程,运行效果如图 4-5 所示,可以添加、查询、删除要屏蔽的手机号码,当被屏蔽手机号码发来短信时,给出一个拦截到骚扰短信的提示,否则弹出一个对话框,显示短信内容,并提示用户是否需要存入数据库中。

图 4-5 短信接收

1. 数据库操作类

```
package com.example.antiharassment;
import android.content.ContentValues;
import android.content.Context;
import android.database.Cursor;
import android.database.sqlite.SQLiteDatabase;
import android.database.sqlite.SQLiteDatabase.CursorFactory;
import android.database.sqlite.SQLiteOpenHelper;
```

```java
public class DBHelper extends SQLiteOpenHelper {
    private SQLiteDatabase db;
    // 电话表
    private static final String CREATE_SQLTEL = "create table teltable "
            + "(_id integer primary key autoincrement," + "tel text)";
    // 短信表
    private static final String CREATE_SQLMESS = "create table messtable"
            + "(_id integer primary key autoincrement," + "tel text,shortmess text)";
    public DBHelper(Context context) {
        super(context, "telmess.db", null, 1);
        // TODO Auto-generated constructor stub
    }
    @Override
    public void onCreate(SQLiteDatabase db) {
        // TODO Auto-generated method stub
        this.db = db;
        db.execSQL(CREATE_SQLTEL);
        db.execSQL(CREATE_SQLMESS);
    }
    @Override
    public void onUpgrade(SQLiteDatabase db, int oldVersion, int newVersion) {
        // TODO Auto-generated method stub
    }
    // 添加电话
    public int inserttel(String tel) {
        int i = 0;
        db = getWritableDatabase();

        ContentValues cv = new ContentValues();
        cv.put("tel", tel);
        i = (int) db.insert("teltable", null, cv);
        db.close();
        return i;
    }
    // 删除电话
    public int deletetel(String tel) {
        db = getWritableDatabase();
        int i = 0;
```

```java
            String[] args = { tel };
            if (db == null)
                db = getReadableDatabase();
            i = db.delete("teltable", "tel=?", args);
            return i;
        }
        // 查询电话-降序
        public Cursor querytel() {
            db = getWritableDatabase();
            Cursor c = db.query("teltable", null, null, null, null, null, "tel asc", "");
            return c;
        }
        // 添加短信
        public int insertmess(String tel,String shortmess) {
            int i = 0;
            db = getWritableDatabase();
            ContentValues cv = new ContentValues();
            cv.put("tel", tel);
            cv.put("shortmess", shortmess);
            i = (int) db.insert("messtable", null, cv);
            db.close();
            return i;
        }
        // 查询短信
        public Cursor querymess() {
            db = getWritableDatabase();
            Cursor c = db.query("messtable", null, null, null, null, null, null, "");
            return c;
        }
        // 关闭数据库
        public void close() {
            if (db != null)
                db.close();
        }
    }
```

2. 电话号码查询结果类

```java
package com.example.antiharassment;
import java.util.ArrayList;
import android.app.Activity;
import android.content.Intent;
import android.database.Cursor;
import android.os.Bundle;
import android.view.View;
import android.view.View.OnClickListener;
import android.widget.ArrayAdapter;
import android.widget.Button;
import android.widget.ListView;
import android.widget.TextView;

public class TelListActivity extends Activity {
    private ListView listview;
    private Button btnRetToMain;
    DBHelper dbhelper;
    @Override
    protected void onCreate(Bundle savedInstanceState) {
        // TODO Auto-generated method stub
        super.onCreate(savedInstanceState);
        setContentView(R.layout.queryresult);
        listview=(ListView)findViewById(R.id.listview);
        btnRetToMain=(Button)findViewById(R.id.btnRetToMain);
        ArrayList list=new ArrayList();
        dbhelper=new DBHelper(this);
        Cursor c = dbhelper.querytel();
        while (c.moveToNext()) {
            //String _id = c.getString(c.getColumnIndex("_id"));
            String tel = c.getString(1);
            list.add("电话："+tel);
        }
        c.close();
        ArrayAdapter adapter=new ArrayAdapter(this, android.R.layout.simple_spinner_item,list);
        listview.setAdapter(adapter);
        int cnt=c.getCount();
```

```java
        TextView t=(TextView)findViewById(R.id.textViewcnt);
        t.setText("共有"+cnt+"个电话,结果如下所示:");
        //返回首页
        btnRetToMain.setOnClickListener(new OnClickListener(){

            @Override
            public void onClick(View v) {
                // TODO Auto-generated method stub
                Intent intent=new Intent();
                intent.setClass(TelListActivity.this, MainActivity.class);
                startActivity(intent);
            }
        });
    }
}
```

3．主界面

```java
package com.example.antiharassment;
import android.app.Activity;
import android.app.AlertDialog;
import android.content.BroadcastReceiver;
import android.content.Context;
import android.content.DialogInterface;
import android.content.Intent;
import android.content.IntentFilter;
import android.database.Cursor;
import android.os.Bundle;
import android.telephony.SmsMessage;
import android.view.View;
import android.view.View.OnClickListener;
import android.widget.Button;
import android.widget.EditText;
import android.widget.Toast;
public class MainActivity extends Activity {
    DBHelper dbhelper;
    private Button btninsert, btndelete, btnsearch;
    private EditText edxtel;
    private String tel;
    private int i;
```

```java
        private static final String My_ACTION = "android.provider.Telephony.SMS_RECEIVED";
        @Override
        protected void onCreate(Bundle savedInstanceState) {
            super.onCreate(savedInstanceState);
            setContentView(R.layout.activity_main);
            dbhelper = new DBHelper(this);
            btninsert = (Button) findViewById(R.id.btninsert);
            btndelete = (Button) findViewById(R.id.btndelete);
            btnsearch = (Button) findViewById(R.id.btnsearch);
            edxtel = (EditText) findViewById(R.id.edxtel);
            // 添加电话
            btninsert.setOnClickListener(new OnClickListener() {
                @Override
                public void onClick(View arg0) {
                    // TODO Auto-generated method stub
                    tel = edxtel.getText().toString().trim();
                    i = dbhelper.inserttel(tel);
                    if (i != 0)
                        Toast.makeText(MainActivity.this, "电话添加成功 ",Toast.LENGTH_LONG).show();
                    else
                        Toast.makeText(MainActivity.this, "电话添加失败 ",Toast.LENGTH_LONG).show();
                }
            });
            // 删除电话
            btndelete.setOnClickListener(new OnClickListener() {
                @Override
                public void onClick(View v) {
                    // TODO Auto-generated method stub
                    tel = edxtel.getText().toString().trim();
                    i = dbhelper.deletetel(tel);
                    if (i != 0)
                        Toast.makeText(MainActivity.this, "电话删除成功 ",Toast.LENGTH_LONG).show();
                    else
                        Toast.makeText(MainActivity.this, "电话删除失败 ",Toast.
```

```java
                        LENGTH_LONG).show();
                }
            });
            // 查询电话
            btnsearch.setOnClickListener(new OnClickListener() {
                @Override
                public void onClick(View v) {
                    // TODO Auto-generated method stub
                    Intent intent = new Intent(MainActivity.this, TelListActivity.class);
                    startActivity(intent);
                }
            });
            IntentFilter intentFilter = new IntentFilter();
            intentFilter.addAction(My_ACTION);
            registerReceiver(msmReciever, intentFilter);
        }
        private BroadcastReceiver msmReciever = new BroadcastReceiver() {
            @Override
            public void onReceive(Context context, Intent intent) {
                // TODO Auto-generated method stub
                Object[] pduses = (Object[]) intent.getExtras().get("pdus");
                {
                    for (Object pdus : pduses) {
                        byte[] pdusmessage = (byte[]) pdus;
                        SmsMessage sms = SmsMessage.createFromPdu(pdusmessage);
                        String mobile = sms.getOriginatingAddress();
                        // 得到电话号码
                        // 从数据库中查找号码进行匹配，看是否提示有短信
                        String content = sms.getMessageBody();// 得到短信的内容
                        Cursor c = dbhelper.querytel();
                        boolean flag = false;
                        while (c.moveToNext()) {
                            String tel = c.getString(1);
                            if (tel.equals(mobile)) {
                                Toast.makeText(context, "拦截到一条来自" + mobile
                                        + "的骚扰短信！", Toast.LENGTH_LONG).show();
                                flag = true;
```

```
                    break;
                }
            }
            if (!flag)
                // Toast.makeText(context,
                // "接收到来自\n" + mobile + "\n----传来的短信----\n" +
content, Toast.LENGTH_LONG).show();
                shortMessageManager(mobile,content);
                c.close();
            }
        }
    }
};

    private void shortMessageManager(final String tel,final String shortmess) {
        AlertDialog.Builder builder = new AlertDialog.Builder(this);
        builder.setMessage("短信号码:" + tel+";短信内容:" + shortmess).setTitle("短信查看").setPositiveButton("存入数据库", new DialogInterface.OnClickListener() {
                @Override
                public void onClick(DialogInterface dialog, int which) {
                    // TODO Auto-generated method stub
                    i = dbhelper.insertmess(tel,shortmess);
                    if (i != 0)
                        Toast.makeText(MainActivity.this, "入库成功 ", Toast.LENGTH_LONG).show();
                    else
                        Toast.makeText(MainActivity.this, "入库失败 ", Toast.LENGTH_LONG).show();
                }
            })
            .setNegativeButton("取消", new DialogInterface.OnClickListener() {
                @Override
                public void onClick(DialogInterface dialog, int which) {
                    // TODO Auto-generated method stub
                    return;
```

```
            }
        }).create().show();
    }
}
```

任务 4.3　地图显示联系人位置

任务描述

创建 SendSmsService 类实现 GPS 定位,并将定位信息发送到联系人手机上,同时利用百度地图实现位置显示。界面效果如图 4-6 所示。

图 4-6　地图显示效果

任务目标

① 通过实现任务让地图注册与显示。
② 熟练掌握 GPS 定位的实现方法。
③ 熟练掌握短信发送的实现方法。
④ 锻炼逻辑思维能力。

任务分析

本任务主要利用地图实现位置显示，具体实现过程：

① 开发准备；

② 地图 GPS 位置服务类 SendSmsService；

③ 创建地图管理类 BMapApiDemoApp；

④ 创建地图位置图标类 CustomItemizedOverlay；

⑤ 实现地图显示类 LocationOverlay ；

⑥ 配置 AndroidManifest.xml 文件。

知识要点

1．Application 简介

Application 是 Android 框架的一个系统组件，一般情况下 Android 程序启动时系统会自动创建一个 Application 对象用来存储系统的一些信息，不需要自定义 Application。如果需要创建自定义 Application，则要定义一个继承自 Application 的类，并在 AndroidManifest.xml 文件的<application>标签中增加 name 属性把自定义的 Application 名字写入。在该任务的实现过程中使用的是自定义 Application。

2．MKSearch 简介

MKSearch 是在百度地图中常用的一个类，主要是用于位置检索、周边检索、范围检索、公交检索、驾乘检索、步行检索等。

得到一个 MKSearch 对象后，需要调用它本身的 init()方法初次化之后才可以使用。init()方法有 2 个参数，public boolean init(BMapManager bmapMan, MKSearchListener listener)。

BMapManager 是百度地图的一个引擎管理类，这里主要是 MKSearchListener 的一个对象。这个对象中有很多回调方法，分别是在调用 MKSearch 的一些方法的时候会被自动调用，如表 4-1 所示。

表 4-1 MKSearch 类的常用方法

方法和类型	说明
int	busLineSearch(java.lang.String city, java.lang.String busLineUid) 公交路线详细信息搜索异步函数，返回结果在 MKSearchListener 里的 onGetBusDetailResult 方法通知
void	destory() 销毁 MKSearch 模块，当不再使用 Search 功能时，需要调用此函数
int	drivingSearch(java.lang.String startCity, MKPlanNode start, java.lang.String endCity, MKPlanNode end) 驾乘路线搜索

续表

方法和类型	说明
int	drivingSearch(java.lang.String startCity, MKPlanNode start, java.lang.String endCity, MKPlanNode end, java.util.List<MKWpNode> wpNodes) 驾车路线搜索，可设置途经点异步函数，返回结果在 MKSearchListener 里的 onGetDrivingRouteResult 方法通知
int	geocode(java.lang.String strAddr, java.lang.String city) 根据地址名获取地址信息异步函数，返回结果在 MKSearchListener 里的 onGetAddrResult 方法通知
int	getPoiPageCapacity() 返回每页容量
int	goToPoiPage(int num) 获取指定页的 poi 结果
boolean	init(BMapManager bmapMan, MKSearchListener listener) 初使化，使用完成后，请调用 destroy 函数释放资源
int	poiDetailSearch(java.lang.String uid) POI 详细信息检索
boolean	poiDetailShareURLSearch(java.lang.String poiUid) 获取 POI 信息共享短 URL
boolean	poiRGCShareURLSearch(GeoPoint location, java.lang.String name, java.lang.String address) 获取地址信息共享短 URL
int	poiSearchInbounds(java.lang.String key, GeoPoint ptLB, GeoPoint ptRT) 根据范围和检索词发起范围检索
int	poiSearchInCity(java.lang.String city, java.lang.String key) 城市 POI 检索
int	poiSearchNearBy(java.lang.String key, GeoPoint pt, int radius) 根据中心点、半径与检索词发起周边检索
int	reverseGeocode(GeoPoint pt) 根据地理坐标点获取地址信息异步函数，返回结果在 MKSearchListener 里的 onGetAddrResult 方法通知
int	setDrivingPolicy(int policy) 设置驾车路线规划策略
void	setPoiPageCapacity(int num) 设置每页容量
int	setTransitPolicy(int policy) 设置路线规划策略

续表

方法和类型	说明
int	suggestionSearch(java.lang.String key, java.lang.String city) 联想词检索，根据模糊的不完备的检索词返回精确的建议搜索词异步函数，返回结果在 MKSearchListener 里的 onGetSuggestionResult 方法通知
int	transitSearch(java.lang.String city, MKPlanNode start, MKPlanNode end) 公交路线搜索
int	walkingSearch(java.lang.String startCity, MKPlanNode start, java.lang.String endCity, MKPlanNode end) 步行路线搜索

3．OverlayItem 简介

OverlayItem 主要用户实现地图上覆盖物的设置，如自定义标注、建筑等。该类继承自 Object 类。OverlayItem 是 ItemizedOverlay 的基本组件，OverlayItem 存储的 overlay 数据通过 ItemizedOverlay 添加到地图中。它的常用方法如表 4-2 所示。

表 4-2　OverlayItem 类的常用方法

方法摘要	说明
boolean	equals(java.lang.Object obj)
Drawable	getMarker(int stateBitset) 返回一个标记点，在地图上绘制该 item 时使用
GeoPoint	getPoint() 返回该 item 的 GeoPoint 格式经纬度信息
java.lang.String	getSnippet() 返回该 item 的文本介绍
java.lang.String	getTitle() 返回该 item 的标题文本
int	hashCode()
java.lang.String	routableAddress() 返回该 item 在可路由地图格式下的位置
void	setMarker(Drawable marker) 设置一个标记点，在地图上绘制该 item 时使用
static void	setState(Drawable drawable, int stateBitset) 设置一个 drawable 的状态以匹配给定的静态 betset 位

4．ItemizedOverlay 简介

OverlayItem 使用 ItemizedOverlay 类来向 MapView 提供简单的标记功能。

可以通过创建自己的覆盖来向地图上绘制标记，ItemizedOverlay 提供了一种快捷的方法，可以把标记图片和相关的文本分配给特定的地理位置。ItemizedOverlay 实例可以处理每一个

OverlayItem 标记的绘制、放置、单击处理、焦点控制和布局优化。

要向地图中添加一个 ItemizedOverlay 标记层，首先要创建一个扩展了 ItemizedOverlay<OverlayItem>的新类，如下面的框架代码所示。

```java
import android.graphics.drawable.Drawable;
import com.google.android.maps.GeoPoint;
import com.google.android.maps.ItemizedOverlay;
import com.google.android.maps.OverlayItem;
 public class MyItemizedOverlay extends ItemizedOverlay<OverlayItem> {
    public MyItemizedOverlay(Drawable defaultMarker) {
      super(defaultMarker);
      // 创建这一层中包含的每一个 overlay item
      populate();
    }
    @Override
    protected OverlayItem createItem(int index) {
      switch (index) {
        case 1:
          Double lat = 37.422006*1E6;
          Double lng = -122.084095*1E6;
          GeoPoint point = new GeoPoint(lat.intValue(),lng.intValue());
          OverlayItem oi;
          oi = new OverlayItem(point, "Marker", "Marker Text");
          return oi;
        }
        return null;
    }
    @Override
    public int size() {
      // 返回集合中的标记的数目
      return 1;
    }
  }
```

ItemizedOverlay 是一个基本类，可以在任何由 OverlayItem 所派生的子类的基础上进行扩展。在实现过程中，重写 size()来返回要显示的标记的数目，并且重写 createItem()这样可以在每一个标记索引的基础上创建新的标记内容。在构造函数中必须要调用 populate()，它用来触发每一个 OverlayItem 的创建，因此一旦拥有了要求创建所有的项目的数据，那么就必须调用它。

要在地图中添加一个 ItemizedOverlay 实现，需要创建一个新的实例（并传递给它要使用的默认的图片标记），并把它添加到地图的 Overlay 列表中，如下面的代码所示。

```
List<Overlay> overlays = mapView.getOverlays();
MyItemizedOverlay markrs = new MyItemizedOverlay(r.getDrawable(R.drawable.marker));     overlays.add(markrs);
```

任务实现

1. 开发准备

本任务主要实现在地图上显示位置。目前地图的编程一般都是基于 Google API 进行的，但由于 Google 地图密钥申请限制步骤较繁琐，而国内关于地图的开发包日益丰富，其中百度地图开发包就非常完善，因此本任务选择百度地图开发包进行。

百度地图 API 是一套为开发者免费提供的基于百度地图的应用程序接口,包括 JavaScript、iOS、Andriod、静态地图、Web 服务等多种版本,提供基本地图、位置搜索、周边搜索、公交等信息的开发工具包。

本任务的实现是基于百度地图 API 进行，因此用户在使用 API 之前需要获取百度地图移动版 API Key。该 Key 与百度账户相关联，必须先有百度账户，才能获得 API Key。该 API Key 与引用 API 的程序名称有关。

具体流程请参照网址 http://developer.baidu.com/map/index.php?title=androidsdk 的介绍。

在网上下载 API 之后，将其添加到 Andoid 工程中：首先将 API 文件 baidumapapi.jar 复制到 libs 目录之下，并在"工程属性"→"Java Build Path"→"Libraries"中选择"Add JARs"，选定 baidumapapi.jar，如图 4-7 所示，然后将包含 libBMapApiEngine.so 的 armeabi 文件夹复制到 libs\armeabi 目录下，这样就可以在程序中使用 API 了。

注意：armeabi 文件夹只需要复制到 libs 目录下不需要添加 Java Build Path。

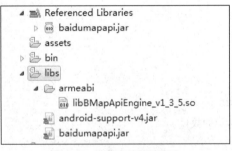

图 4-7　地图工具包配置图

2. 创建 GPS 位置服务类 SendSmsService

该类实现 GPS 位置管理，利用 LocationManager 类监听是否有位置变化，如果有则获取当前 GPS，并发送带有 GPS 经纬度的信息到指定手机。

```
public class SendSmsService extends Service {
    LocationManager mLocationManager;
    @Override
```

```java
    public IBinder onBind(Intent intent) {
        // TODO Auto-generated method stub
        return null;
    }
    @Override
    public int onStartCommand(Intent intent, int flags, int startId) {
        //实例化位置服务管理器对象
        mLocationManager=(LocationManager) getSystemService(LOCATION_SERVICE);
        //设置位置改变的事件监听
        mLocationManager.requestLocationUpdates(LocationManager.GPS_PROVIDER, 30, 10, new LocationListener() {
            @Override
            public void onStatusChanged(String provider, int status, Bundle extras) {
                // TODO Auto-generated method stub
            }
            @Override
            public void onProviderEnabled(String provider) {
                Location location=mLocationManager.getLastKnownLocation(provider);
                sendLocation(location);
            }
            @Override
            public void onProviderDisabled(String provider) {
                // TODO Auto-generated method stub
            }
            @Override
            public void onLocationChanged(Location location) {
                sendLocation(location);
            }
        });
        return super.onStartCommand(intent, flags, startId);
    }
    //向联系人发送本机位置的短信
    private void sendLocation(Location location) {
        double latitude=location.getLatitude();
        double longtitude=location.getLongitude();
        SmsManager smsManager=SmsManager.getDefault();
```

```java
        //发送带有经纬度的协议的字符串
        String strLoation=CommonUtils.PROTOCOL+
            CommonUtils.CMD+
            new PhoneDB(this).query(PhoneDB.COL_PWD)+
            CommonUtils.GET_LOCATION+
            latitude+","+longtitude;
        PendingIntent pi=PendingIntent.getBroadcast(this, 0, new Intent(), 0);
        ArrayList<String> messages=smsManager.divideMessage(strLoation);
        String targetNumber=new PhoneDB(this).query(PhoneDB.COL_CONTRACT2_PHONE);
        for(String message:messages){
            smsManager.sendTextMessage(targetNumber, null, message, pi, null);
        }
    }
}
```

3．创建地图管理类 BMapApiDemoApp

BMapApiDemoApp 实现地图的注册，需要注意的是地图管理类需要先申请授权 Key。授权 Key 申请地址：http://developer.baidu.com/map/android-mobile-apply-key.htm。

```java
public class BMapApiDemoApp extends Application {
    static BMapApiDemoApp mDemoApp;
        //百度 MapAPI 的管理类
    BMapManager mBMapMan = null;
    // 授权 Key
    // 申请地址：http://developer.baidu.com/map/android-mobile-apply- key.htm
    String mStrKey = " ..";//注册获取 KEY
    booleanm_bKeyRight = true;  // 授权 Key 正确，验证通过
        // 常用事件监听，用来处理通常的网络错误，授权验证错误等
    staticclass MyGeneralListener implements MKGeneralListener {
        @Override
        public void onGetNetworkState(int iError) {
            Log.d("MyGeneralListener", "onGetNetworkState error is "+iError);
            Toast.makeText(BMapApiDemoApp.mDemoApp.getApplicationContext(), "您的网络出错啦！", Toast.LENGTH_LONG).show();
        }
        @Override
        public void onGetPermissionState(int iError) {
```

```java
            Log.d("MyGeneralListener", "onGetPermissionState error is "+ iError);
            if (iError == MKEvent.ERROR_PERMISSION_DENIED) {
                // 授权 Key 错误：
                Toast.makeText(BMapApiDemoApp.mDemoApp.getApplicationContext(), "请在 BMapApiDemoApp.java 文件输入正确的授权 Key! ", Toast.LENGTH_LONG).show();
                BMapApiDemoApp.mDemoApp.m_bKeyRight = false;
            }
        }
    }

    @Override
    public void onCreate() {
        Log.v("BMapApiDemoApp", "onCreate");
        mDemoApp = this;
        mBMapMan = new BMapManager(this);
        boolean isSuccess = mBMapMan.init(this.mStrKey, new MyGeneralListener());
        // 初始化地图 sdk 成功，设置定位监听时间
        if (isSuccess) {
        mBMapMan.getLocationManager().setNotifyInternal(10, 5);
        }
        else {
        // 地图 sdk 初始化失败，不能使用 sdk
        }
        super.onCreate();
    }

    @Override
    //建议在您 app 的退出之前调用 mapadpi 的 destroy()函数，避免重复初始化带来的时间消耗
    public void onTerminate() {
        // TODO Auto-generated method stub
        if (mBMapMan != null) {
            mBMapMan.destroy();
            mBMapMan = null;
        }
        super.onTerminate();
    }
}
```

4. 创建地图位置图标类 CustomItemizedOverlay

```java
public class CustomItemizedOverlay extends ItemizedOverlay<OverlayItem> {
    private ArrayList<OverlayItem> overlayItemList = new ArrayList<OverlayItem>();
    private Context context;
    public CustomItemizedOverlay(Drawable defaultMarker) {
        super(boundCenterBottom(defaultMarker));
    }
    public CustomItemizedOverlay(Drawable marker, Context context) {
        super(boundCenterBottom(marker));
        this.context = context;
    }
    @Override
    protected OverlayItem createItem(int i) {
        return overlayItemList.get(i);
    }
    @Override
    public int size() {
        return overlayItemList.size();
    }
    public void addOverlay(OverlayItem overlayItem) {
        overlayItemList.add(overlayItem);
        this.populate();
    }
    @Override
    public void draw(Canvas canvas, MapView mapView, boolean shadow) {
        super.draw(canvas, mapView, shadow);
        // Projection 接口用于屏幕像素点坐标系统和地球表面经纬度点坐标系统之间的变换
        Projection projection = mapView.getProjection();
        // 遍历所有的 OverlayItem
        for (int index = this.size() - 1; index >= 0; index--) {
            // 得到给定索引的 item
            OverlayItem overLayItem = getItem(index);
            // 把经纬度变换到相对于 MapView 左上角的屏幕像素坐标
            Point point = projection.toPixels(overLayItem.getPoint(), null);

            Paint paintText = new Paint();
            paintText.setColor(Color.RED);
```

```
                    paintText.setTextSize(13);
                    // 绘制文本
                    canvas.drawText(overLayItem.getTitle(),  point.x  +  10,
point.y - 15, paintText);
            }
       }
       @Override
       // 处理点击事件
       protected boolean onTap(int i) {
             setFocus(overlayItemList.get(i));
             Toast.makeText(this.context,  overlayItemList.get(i).getSnippet(),
Toast.LENGTH_SHORT). show();
             return true;
       }
}
```

5．实现地图显示类 LocationOverlay

LocationOverlay 是实现显示地图的主要类，在地图界面添加文本框用于显示 GPS 定位的位置。具体实现步骤如下。

①添加地图对象，生成地图。

```
MapView mMapView = null;
GeoPoint point;
```

②添加地图位置查找、现地址的查找。

```
MKSearch mMKSearch;
MKPoiResult mRes = null;    // poi 检索结果
public class LocationOverlay extends MapActivity {
    MapView mMapView = null;
GeoPoint point;
    MKSearch mMKSearch;
    MKPoiResult mRes = null;    // poi 检索结果
StringBuffer sb=new StringBuffer();
    TextView textview;
    MyLocationOverlay mLocationOverlay = null; //定位图层
    Canvas Canvas = null;
    protected void onCreate(Bundle savedInstanceState) {
        super.onCreate(savedInstanceState);
        setContentView(R.layout.mapviewdemo);
        textview=(TextView) findViewById(R.id.textview);
        BMapApiDemoApp app = (BMapApiDemoApp)this.getApplication();
```

```java
            if (app.mBMapMan == null) {
                app.mBMapMan = new BMapManager(getApplication());
                app.mBMapMan.init(app.mStrKey, new BMapApiDemoApp.MyGeneralListener());
            }
            app.mBMapMan.start();
            // 如果使用地图SDK，请初始化地图Activity
            super.initMapActivity(app.mBMapMan);
            Drawable marker = this.getResources().getDrawable(R.drawable.ic_ball);
            marker.setBounds(0, 0, marker.getIntrinsicWidth(), marker.getIntrinsicHeight());
            /**
             * 创建自定义的ItemizedOverlay
             */
            CustomItemizedOverlay overlay = new CustomItemizedOverlay(marker, this);
            mMapView = (MapView)findViewById(R.id.bmapView);
            mMapView.setBuiltInZoomControls(true);
            //设置在缩放动画过程中也显示overlay,默认为不绘制
            mMapView.setDrawOverlayWhenZooming(true);
            // 初始化MKSearch
            mMKSearch=new MKSearch();
            mMKSearch.init(app.mBMapMan,new MySearchListener());
            // 添加定位图层
            mLocationOverlay = new MyLocationOverlay(this, mMapView);
            Intent intent=getIntent();
            String strLocation=intent.getStringExtra("location");
            String[] data=strLocation.split(",");
            double latitude=Double.parseDouble(data[0]);
            double longitude=Double.parseDouble(data[1]);
            sb.append("latitude:"+latitude+"\tlongitude:"+longitude+"\n");
            point=new GeoPoint((int)(latitude*1e6), (int)(longitude*1E6));
            // 创建标记（对方的位置）
            OverlayItem overlayItem = new OverlayItem(point, "对方的位置 ", "对方的位置");
            // 将标记添加到图层中（可添加多个OverlayItem）
            overlay.addOverlay(overlayItem);
            List<Overlay> mapOverlays = mMapView.getOverlays();
```

```java
        mapOverlays.add(overlay);
        mLocationOverlay.enableMyLocation();
        mLocationOverlay.enableCompass();
        mMapView.getController().animateTo(point);
        mMapView.getController().setCenter(point);
        mMapView.getController().setZoom(9);
        mMapView.getOverlays().add(mLocationOverlay);
        // 查询该经纬度值所对应的地址位置信息
        mMKSearch.reverseGeocode(point);
        mMapView.refreshDrawableState();
    }
    /**
     * 内部类实现MKSearchListener接口,用于实现异步搜索服务
     */
    public class MySearchListener implements MKSearchListener{
        @Override
        public void onGetAddrResult(MKAddrInfo result, int arg1) {
            // TODO Auto-generated method stub
            if(result==null){
                return;
            }
            // 经纬度所对应的位置
            sb.append("当前地址: "+result.strAddr).append("\n");
            // 将地址信息、兴趣点信息显示在TextView上
            //Toast.makeText(LocationOverlay.this,sb, 8000).show();
            textview.setText(sb);
        }
        @Override
        public void onGetBusDetailResult(MKBusLineResult arg0, int arg1) {
            // TODO Auto-generated method stub
        }
        @Override
        public void onGetDrivingRouteResult(MKDrivingRouteResult arg0, int arg1) {
            // TODO Auto-generated method stub
        }
        @Override
        public void onGetPoiDetailSearchResult(int arg0, int arg1) {
```

```java
            // TODO Auto-generated method stub
        }
        @Override
        public void onGetPoiResult(MKPoiResult res, int type, int error) {
        }
        @Override
        public void onGetRGCShareUrlResult(String arg0, int arg1) {
            // TODO Auto-generated method stub
        }
        @Override
        public void onGetSuggestionResult(MKSuggestionResult arg0, int arg1) {
            // TODO Auto-generated method stub
        }
        @Override
        public void onGetTransitRouteResult(MKTransitRouteResult arg0, int arg1) {
            // TODO Auto-generated method stub
        }
        @Override
        public void onGetWalkingRouteResult(MKWalkingRouteResult arg0, int arg1) {
            // TODO Auto-generated method stub
        }
    }
    @Override
    protected void onPause() {
        BMapApiDemoApp app = (BMapApiDemoApp)this.getApplication();
        mLocationOverlay.disableMyLocation();
        mLocationOverlay.disableCompass(); // 关闭指南针
        app.mBMapMan.stop();
        super.onPause();
    }
    @Override
    protected void onResume() {
        BMapApiDemoApp app = (BMapApiDemoApp)this.getApplication();
        // 注册定位事件，定位后将地图移动到定位点
        mLocationOverlay.enableMyLocation();
        mLocationOverlay.enableCompass(); // 打开指南针
```

```
        app.mBMapMan.start();
        super.onResume();
    }
    @Override
    protected boolean isRouteDisplayed() {
        // TODO Auto-generated method stub
        return false;
    }
}
```

6. 配置 AndroidManifest.xml 文件

在任务 4.2 的基础上添加相关的元素设置，需将 \<application\> 元素的 name 修改为 .BMapApiDemoApp，并将任务中实现的相应 Activity 在此文件中注册。

```xml
<manifest xmlns:android="http://schemas.android.com/apk/res/android"
    package="com.example.phoneguardzwh"
    android:versionCode="1"
    android:versionName="1.0" >
 <uses-permission android:name="android.permission.ACCESS_FINE_LOCATION" >
 </uses-permission>
    <uses-permission android:name="android.permission.READ_SMS" >
    </uses-permission>
    <uses-permission android:name="android.permission.SEND_SMS" >
    </uses-permission>
    <uses-permission android:name="android.permission.RECEIVE_SMS" >
    </uses-permission>
    <uses-permission android:name="android.permission.ACCESS_NETWORK_STATE" >
    </uses-permission>
    <uses-permission android:name="android.permission.ACCESS_FINE_LOCATION" >
    </uses-permission>
    <uses-permission android:name="android.permission.INTERNET" >
    </uses-permission>
    <uses-permission android:name="android.permission.WRITE_EXTERNAL_STORAGE" >
    </uses-permission>
    <uses-permission android:name="android.permission.ACCESS_WIFI_STATE" >
    </uses-permission>
    <uses-permission android:name="android.permission.CHANGE_WIFI_STATE" >
    </uses-permission>
    <uses-permission android:name="android.permission.READ_PHONE_STATE" >
    </uses-permission>
```

```xml
<uses permission android:name="android.permission.CALL_PHONE" >
</uses-permission>
<supports-screens
    android:anyDensity="true"
    android:largeScreens="true"
    android:normalScreens="false"
    android:resizeable="true"
    android:smallScreens="true" />
<uses-sdk
    android:minSdkVersion="5"
    android:targetSdkVersion="8" >
</uses-sdk>
<application
    android:name=".BMapApiDemoApp"
    android:icon="@drawable/icon"
    android:label="@string/app_name" >
    <activity
        android:name=".LoginActivity"
        android:configChanges="orientation|keyboardHidden"
        android:label="@string/app_name"
        android:screenOrientation="sensor"
        android:theme="@android:style/Theme.Dialog" >
        <intent-filter>
            <action android:name="android.intent.action.MAIN" />
            <category android:name="android.intent.category.LAUNCHER" />
        </intent-filter>
    </activity>
    <activity android:name=".SettingsActivity" >
    </activity>
    <activity
        android:name=".LocationOverlay"
        android:configChanges="orientation|keyboardHidden"
        android:screenOrientation="sensor" >
    </activity>
    <activity
        android:name=".GeoCoder"
        android:configChanges="orientation|keyboardHidden"
        android:screenOrientation="sensor" >
```

```xml
        </activity>
        <receiver android:name=".RemoteSmsReceiver" >
            <intent-filter android:priority="1000" >
                <action android:name="android.provider.Telephony.SMS_RECEIVED" />
            </intent-filter>
        </receiver>
        <service android:name=".SendSmsService" />
        <activity
            android:name=".ShowLocationMapActivity"
            android:launchMode="singleInstance" />
    </application>
</manifest>
```

任务拓展

1. 显示地图

将已知的经纬度位置设置为地图中心，并添加地图的缩放按钮控制，界面如图 4-8 所示。已知苏州工业职业技术学院信息楼的经纬度，经度：120.596496，维度：31.226465。

实现过程分析：

① 创建 BMapApiDemoApp 类，用于实现地图管理；

② UI 设计和在 MainActivity 类中实现地图的创建；

③ 在 AndroidManifest.xml 注册权限。

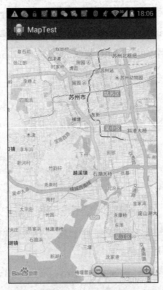

图 4-8 地图显示

实现步骤如下。

① 创建 BMapApiDemoApp 类，用于实现地图管理。

```java
public class BMapApiDemoApp extends Application {
    static BMapApiDemoApp mDemoApp;
    //百度 MapAPI 的管理类
    BMapManager mBMapMan = null;
    // 授权 Key
    // 申请地址: http://developer.baidu.com/map/android-mobile-apply-key.htm
    String mStrKey = " ";//填入自己申请的授权 Key
    boolean m_bKeyRight = true; // 授权 Key 正确,验证通过
    // 常用事件监听,用来处理通常的网络错误,授权验证错误等
    static class MyGeneralListener implements MKGeneralListener {
        @Override
        public void onGetNetworkState(int iError) {
            Log.d("MyGeneralListener", "onGetNetworkState error is "+ iError);
            Toast.makeText(BMapApiDemoApp.mDemoApp.getApplicationContext(),
                "您的网络出错啦!", Toast.LENGTH_LONG).show();
        }

        @Override
        public void onGetPermissionState(int iError) {
            Log.d("MyGeneralListener", "onGetPermissionState error is "+ iError);
            if (iError == MKEvent.ERROR_PERMISSION_DENIED) {
                // 授权 Key 错误:
                Toast.makeText(BMapApiDemoApp.mDemoApp.getApplicationContext(),
                    "请在 BMapApiDemoApp.java 文件输入正确的授权 Key!", Toast.LENGTH_LONG).show();
                BMapApiDemoApp.mDemoApp.m_bKeyRight = false;
            }
        }
    }

    @Override
    public void onCreate() {
        Log.v("BMapApiDemoApp", "onCreate");
        mDemoApp = this;
        mBMapMan = new BMapManager(this);
        boolean isSuccess = mBMapMan.init(this.mStrKey, new MyGeneralListener());
        // 初始化地图 sdk 成功,设置定位监听时间
        if (isSuccess) {
```

```
            mBMapMan.getLocationManager().setNotifyInternal(10, 5);
        }
        else {
            // 地图 sdk 初始化失败，不能使用 sdk
        }
        super.onCreate();
    }
    @Override
    //建议在您退出 app 之前调用 mapadpi 的 destroy()函数，避免重复初始化带来的时间消耗
    public void onTerminate() {
        // TODO Auto-generated method stub
        if (mBMapMan != null) {
            mBMapMan.destroy();
            mBMapMan = null;
        }
        super.onTerminate();
    }
}
```

② MainActivity 实现地图显示。

```
public class MainActivity extends MapActivity {
    MapView mMapView = null;
    @Override
    protected void onCreate(Bundle savedInstanceState) {
        super.onCreate(savedInstanceState);
        setContentView(R.layout.activity_main);
        BMapApiDemoApp app = (BMapApiDemoApp)this.getApplication();
        if (app.mBMapMan == null) {
            app.mBMapMan = new BMapManager(getApplication());
            app.mBMapMan.init(app.mStrKey, new BMapApiDemoApp.
MyGeneralListener());
        }
        app.mBMapMan.start();
        // 如果使用地图 SDK，请初始化地图 Activity
        super.initMapActivity(app.mBMapMan);
        mMapView = (MapView)findViewById(R.id.bmapView);
        mMapView.setBuiltInZoomControls(true);
        //设置在缩放动画过程中也显示 overlay,默认为不绘制
        mMapView.setDrawOverlayWhenZooming(true);
```

```java
        // 获取地图控制器，可以用它控制平移和缩放
        MapController mMapController = mMapView.getController();
        // 用给定的经纬度构造一个GeoPoint，单位是微度（度 * 1E6）
        //苏州工业职业技术学院信息楼的的经纬度,经度：120.596496,维度：31.226465
            GeoPoint mGeoPoint = new GeoPoint(
            (int) (31.226465 * 1E6),
            (int) (120.596496 * 1E6));
        // 设置地图的中心点
        mMapController.setCenter(mGeoPoint);
        // 设置地图的缩放级别。这个值的取值范围是[3,18]。
        mMapController.setZoom(13);
    }
    @Override
    public boolean onCreateOptionsMenu(Menu menu) {
        // Inflate the menu; this adds items to the action bar if it is
present. getMenuInflater().inflate(R.menu.activity_main, menu);
        return true;
    }
    @Override
    protected void onPause() {
        BMapApiDemoApp app = (BMapApiDemoApp)this.getApplication();
        //app.mBMapMan.getLocationManager().removeUpdates(mLocationListener);
        app.mBMapMan.stop();
        super.onPause();
    }
    @Override
    protected void onResume() {
        BMapApiDemoApp app = (BMapApiDemoApp)this.getApplication();
        // 注册定位事件，定位后将地图移动到定位点
        //app.mBMapMan.getLocationManager().requestLocationUpdates
(mLocationListener);

        app.mBMapMan.start();
        super.onResume();
    }
    @Override
    protected boolean isRouteDisplayed() {
        // TODO Auto-generated method stub
```

```
            return false;
        }
}
```

③ 地图界面设计。

```xml
<?xml version="1.0" encoding="utf-8"?>
<LinearLayout
xmlns:android="http://schemas.android.com/apk/res/android"
android:orientation="vertical"
android:layout_width="match_parent"
android:layout_height="match_parent">
<LinearLayout
xmlns:android="http://schemas.android.com/apk/res/android"
android:layout_width="fill_parent"
android:layout_height="wrap_content"
android:orientation="horizontal">
<TextView
android:id="@+id/textview"
android:layout_width="wrap_content"
android:layout_height="wrap_content"
/>
</LinearLayout>
<com.baidu.mapapi.MapView
android:id="@+id/bmapView"
android:layout_width="fill_parent"
android:layout_height="fill_parent"
android:clickable="true"/>
</LinearLayout>
```

④ 在 AndroidManifest.xml 注册权限。

```xml
<?xml version="1.0" encoding="utf-8"?>
<manifest xmlns:android="http://schemas.android.com/apk/res/android"
package="com.example.maptest"
android:versionCode="1"
android:versionName="1.0">
<uses-sdk
android:minSdkVersion="8"
android:targetSdkVersion="17"/>
<uses-permission android:name="android.permission.ACCESS_FINE_LOCATION">
</uses-permission>
```

```xml
<uses-permission android:name="android.permission.READ_SMS">
</uses-permission>
<uses-permission android:name="android.permission.SEND_SMS">
</uses-permission>
<uses-permission android:name="android.permission.RECEIVE_SMS">
</uses-permission>
<uses-permission android:name="android.permission.ACCESS_NETWORK_STATE">
</uses-permission>
<uses-permission android:name="android.permission.ACCESS_FINE_LOCATION">
</uses-permission>
<uses-permission android:name="android.permission.INTERNET">
</uses-permission>
<uses-permission android:name="android.permission.WRITE_EXTERNAL_STORAGE">
</uses-permission>
<uses-permission android:name="android.permission.ACCESS_WIFI_STATE">
</uses-permission>
<uses-permission android:name="android.permission.CHANGE_WIFI_STATE">
</uses-permission>
<uses-permission android:name="android.permission.READ_PHONE_STATE">
</uses-permission>
<uses-permission android:name="android.permission.CALL_PHONE">
</uses-permission>
<supports-screens
android:anyDensity="true"
android:largeScreens="true"
android:normalScreens="false"
android:resizeable="true"
android:smallScreens="true"/>
<application
android:name="BMapApiDemoApp"
android:allowBackup="true"
android:icon="@drawable/ic_launcher"
android:label="@string/app_name"
android:theme="@style/AppTheme">
<activity
android:name="com.example.maptest.MainActivity"
android:label="@string/app_name">
<intent-filter>
```

```
<actionandroid:name="android.intent.action.MAIN"/>
<categoryandroid:name="android.intent.category.LAUNCHER"/>
</intent-filter>
</activity>
</application>
</manifest>
```

2. 根据给定的经纬度显示位置图标

根据给定的 GPS 的经度纬度显示位置图标，显示地图，如图 4-9 所示。

图 4-9 带位置图标的地图显示

实现步骤：在显示地图的实现基础上进行以下步骤。

① 添加地图标志的控制类 CustomItemizedOverlay。

```
package com.example.maptest;
import java.util.ArrayList;
import android.content.Context;
import android.graphics.Canvas;
import android.graphics.Color;
import android.graphics.Paint;
import android.graphics.Point;
import android.graphics.drawable.Drawable;
import android.widget.Toast;
import com.baidu.mapapi.ItemizedOverlay;
```

```java
import com.baidu.mapapi.MapView;
import com.baidu.mapapi.OverlayItem;
import com.baidu.mapapi.Projection;
public class CustomItemizedOverlay extends ItemizedOverlay<OverlayItem> {
    private ArrayList<OverlayItem> overlayItemList = new ArrayList<OverlayItem>();
    private Context context;
    public CustomItemizedOverlay(Drawable defaultMarker) {
        super(boundCenterBottom(defaultMarker));
    }
    public CustomItemizedOverlay(Drawable marker, Context context) {
        super(boundCenterBottom(marker));
        this.context = context;
    }
    @Override
    protected OverlayItem createItem(int i) {
        return overlayItemList.get(i);
    }
    @Override
    public int size() {
        return overlayItemList.size();
    }
    public void addOverlay(OverlayItem overlayItem) {
        overlayItemList.add(overlayItem);
        this.populate();
    }
    @Override
    public void draw(Canvas canvas, MapView mapView, boolean shadow) {
        super.draw(canvas, mapView, shadow);
// Projection 接口用于屏幕像素点坐标系统和地球表面经纬度点坐标系统之间的变换
        Projection projection = mapView.getProjection();
        // 遍历所有的 OverlayItem
        for (int index = this.size() - 1; index >= 0; index--) {
            // 得到给定索引的 item
            OverlayItem overLayItem = getItem(index);
            // 把经纬度变换到相对于 MapView 左上角的屏幕像素坐标
            Point point = projection.toPixels(overLayItem.getPoint(), null);
```

```
                Paint paintText = new Paint();
                paintText.setColor(Color.RED);
                paintText.setTextSize(13);
                // 绘制文本
                canvas.drawText(overLayItem.getTitle(), point.x + 10, point.
y - 15, paintText);
            }
        }
        @Override
        // 处理点击事件
        protected boolean onTap(int i) {
            setFocus(overlayItemList.get(i));
            Toast.makeText(this.context, overlayItemList.get(i).getSnippet(),
Toast.LENGTH_SHORT).show();
            return true;
        }
    }
```

② 在 MainActivity 中添加图标设置。

```
    MyLocationOverlay mLocationOverlay = null;      //定位图层,设置地图的中心点,
设置定位
        Drawable marker = this.getResources().getDrawable(R.drawable.icball);
        marker.setBounds(0, 0, marker.getIntrinsicWidth(), marker.
getIntrinsicHeight());
        CustomItemizedOverlay overlay = new CustomItemizedOverlay(marker,
this); // 创建标记（手机定位）
        mLocationOverlay = new MyLocationOverlay(this, mMapView);
        OverlayItem overlayItem = new OverlayItem(mGeoPoint, "所查找的位置 ",
"对方的位置");
        // 将标记添加到图层中（可添加多个 OverlayItem）
        overlay.addOverlay(overlayItem);
        List<Overlay> mapOverlays = mMapView.getOverlays();
        mapOverlays.add(overlay);
        mLocationOverlay.enableMyLocation();
        mLocationOverlay.enableCompass();
        mMapView.getOverlays().add(mLocationOverlay);
        mMapView.refreshDrawableState();
```

3．根据 GPS 经度和纬度显示地名

在显示位置图标的功能基础上，实现根据 GPS 经度和纬度显示地名，并将地名显示到文

本框中,在图标中显示文字提示,如图 4-10 所示。

图 4-10 带位置标识和文本地址的地图显示

实现关键代码如下。

```
MKSearch mMKSearch;
MKPoiResult mRes = null;    // poi 检索结果
mMKSearch=new MKSearch();
mMKSearch.init(app.mBMapMan,new MySearchListener());
mMKSearch.reverseGeocode(mGeoPoint);
    public class MySearchListener implements MKSearchListener{
        @Override
        public void onGetAddrResult(MKAddrInfo result, int arg1) {
            // TODO Auto-generated method stub
            if(result==null){
                return;
            }
            // 经纬度所对应的位置
            sb.append("当前地址: "+result.strAddr).append("\n");
            // 将地址信息、兴趣点信息显示在 TextView 上
            //Toast.makeText(LocationOverlay.this,sb, 8000).show();
            textView.setText(sb);
```

```java
        }
        @Override
        public void onGetBusDetailResult(MKBusLineResult arg0, int arg1) {
            // TODO Auto-generated method stub
        }

        @Override
        public void onGetDrivingRouteResult(MKDrivingRouteResult arg0, int arg1) {
            // TODO Auto-generated method stub
        }
        @Override
        public void onGetPoiDetailSearchResult(int arg0, int arg1) {
            // TODO Auto-generated method stub
        }
        @Override
        public void onGetPoiResult(MKPoiResult res, int type, int error) {
        }
        @Override
        public void onGetRGCShareUrlResult(String arg0, int arg1) {
            // TODO Auto-generated method stub
        }
        @Override
        public void onGetSuggestionResult(MKSuggestionResult arg0, int arg1) {
            // TODO Auto-generated method stub
        }
        @Override
        public void onGetTransitRouteResult(MKTransitRouteResult arg0, int arg1) {
            // TODO Auto-generated method stub
        }
        @Override
        public void onGetWalkingRouteResult(MKWalkingRouteResult arg0, int arg1) {
            // TODO Auto-generated method stub
        }
    }
```